Building Safety Commitment

Daniel J. Moran, Ph.D., BCBA-D

Daniel.Moran@qualitysafetyedge.com

www.buildingsafetycommitment.com

Valued Living Books, Inc.
1415 Maple Road
Joliet, IL 60432
www.valuedlivingbooks.com

To my father,

Dan Moran – Electrician

Praise for Building Safety Commitment

"Dr. Moran's ideas have dramatically changed our company. I believe that Dr. Moran has developed a concept and a teaching method that will revolutionize the construction industry."
- Steve Turner, *Founder & CEO*, Corrosion Monitoring Services

"I have been through a lot of safety training in my 22-year career and this really hits home. I especially like the aspect of identifying what an individual values and then put that to use in the work setting. If someone uses what they value as a tool there is no limit on what you can achieve in safety."
- Ed Bahr, *Safety Director*, Boise Inc., Alabama Operations

"I have heard Dr. D.J. Moran speak several times in the last year. I believe he is on the cutting edge of changing the way we work to increase safety behaviors by focusing personal commitment. His success in impacting behavior outside of industrial safety has direct application in the field of worker safety. I predict that in 15-20 years, we will be discussing Dr. Moran's work as a step change in improving behavior-based safety."
- Ron Truelove, *28-Year Environmental, Health, and Safety Professional*, Environmental, Health, and Safety Manager for an Oil and Gas Company

"Dr. D.J. has clearly distinguished himself in the field of safety. Workers, managers, and leadership can all benefit through the practical application of Dr. Moran's safety training."
- Deb Smith, *Director of Quality and Training*, The Horton Group

"We live in a world of made things, but we sometimes forget the makers. In fact, part of the art of the craftsman is to leave behind a product that is pleasing to the eye and functional to the hand, but where the sweat, toil, and risk of the making are inapparent.
D.J. Moran knows this world behind the world and has heart for the makers of things. *Building Safety Commitment* brings solid science to the workplace. This book will keep the people that make the world we live in safer and that is a job worth doing."
- Kelly G. Wilson, Author of *Wisdom to Know the Difference*

Dear Reader,

This book is written for front-line workers as the primary audience. Of course, *Building Safety Commitment* will be helpful to safety professionals, managers, supervisors, and company leaders, as well. A company-wide implementation of the concepts and worksheets in this book will have an optimal influence on safety.

My hope is that men and women working in dynamic environments will use the tools in this book to stay safe in the workplace. Also keep in mind that the concepts in *Building Safety Commitment* can be used throughout the day - even when you leave the jobsite - because personal safety is important no matter where you go.

Enjoy and be well,

D.J. Moran

Table of Contents

Foreword

It is my pleasure to have Dr. Daniel J. Moran as a team member at my company, Quality Safety Edge, and I am very interested in seeing the ideas in *Building Safety Commitment* become more widespread in the safety industry. When I founded Quality Safety Edge over two decades ago, I had a mission to assist world-class companies by implementing behavior-based safety processes that make measurable, lasting improvements in operations and financial performance. My colleagues and I have over 20 years of confirmed results in a variety of industries, and we have stayed committed to the application of science to improving safety. My book, *The Values-Based Safety Process*, outlines the design of unique processes for the specific needs of organizations that entrust us with their safety goals.

Quality Safety Edge is always looking for reliable ways to improve our results and create greater satisfaction for our clients. When I met Dr. Daniel J. Moran (who prefers to be called "D.J.") and learned more about his background and career goals, I recognized that he would be a sensational fit for where I envisioned taking the company. He is a well-trained Ph.D. psychologist who is board certified in the science that is at the very foundation of behavior-based safety. More importantly, he had a unique idea that I thought would make an important contribution to the science of safety: he aims to use behavioral science to help people increase their ability to commit to safer actions.

After several decades in this business, I continue to see one significant concern that safety training and safety equipment is not answering correctly, and D.J.'s ideas address this problem. No matter how well a company writes a safety policy, and no matter how much money they spend on safety equipment, those endeavors will not be helpful unless the employees and leaders are committed to correctly following the policy and using the equipment. D.J. helps people become more committed to their work, their safety, their actions, and their values. He is recognized globally as expert in a field called Acceptance and Commitment Training (ACTraining), has written one of the leading texts on ACTraining, and teaches these techniques to professionals around the world. What struck me as most remarkable was that ACTraining encourages people to make values-based decisions, which was a perfect fit for the Values-Based Safety® approach we use at Quality Safety Edge.

When I invited D.J. to dinner so that he could meet the other professionals at QSE, I was astonished when our waiter asked him for his autograph. I wasn't aware of it at the time, but D.J. consults with Discovery Studios on television shows focusing on mental health issues. Our waiter said that he recognized D.J. from a show he did called *Hoarding: Buried Alive*. When D.J. signed the napkin for the waiter, he addressed it to the young man and then wrote: "If you save this autographed napkin, you might be on the road to becoming a hoarder. Throw it out!" I liked D.J.'s humor and humility, and I am glad he is bringing his expertise to my team.

In the world of safety, D.J.'s techniques help people develop committed actions that demonstrate their true value for the safety of themselves and their associates, both at home and at work. Whether you are reading through this book on your own, or with a team of other folks on your job, this book provides a framework towards a fresh approach to safety. It gives you a set of tools that can help guide you through the process of clarifying your values.

In addition, the tools and skills you will learn from this book with help you deal with distractions, lack of motivation, feeling rushed, and other obstacles that may be prompting unsafe behavior or sending the wrong message to others in your environment. With these new tools, you can work to increase the frequency and consistency of values-based actions. But notice that word "work" in the last sentence. Work is a prerequisite to getting value from this book. You have to do the work with the tools he is providing if you want to see the payoff of building safety commitment. I am sure you will benefit from the investment of your time and labor.

Best wishes,
Terry E. McSween, Ph.D.
Author, *The Values-Based Safety Process*
CEO, Quality Safety Edge

Acknowledgements

Building Safety Commitment began when I was invited to improve the safety culture of a world-class organization operating in dangerous work environments. Steve Turner, the CEO of Corrosion Monitoring Services, and Brian Schifler, veteran supervisor of the company, had a group of employees who needed a push to help them maintain their commitment to safety. We collaborated on integrating effective safety processes into the company's work culture, and it was met with wonderful success. Once their employees improved their safety commitments, Brian and Steve urged me to go do the same thing for other companies, and I have gone on to help thousands of workers learn how to strengthen their commitments. I am grateful for Brian and Steve's encouragement.

Quality Safety Edge (QSE) saw the importance of bringing commitment-based interventions to their *Values-Based Safety*[®] approach, and fortunately for me, brought me into their consulting family. I am continually amazed at the work that QSE does around the world, and the opportunities I have been given because I am part of the team. I especially appreciate Terry McSween for writing the foreword, Jerry Pounds for being my coach and mentor, and Angelica Grindle for her comments and critiques during the development of my ideas. Many thanks to Judith Stowe, Beth Foate, Grainne Matthews, Nathan Crutchfield, and Bob Foxworthy for making me feel part of the QSE team. I'd also like to thank Christian Ingle for helping me get my feet wet in international consulting.

Randy Burgess was instrumental to me getting my thoughts and experiences down into a book format, and I am very grateful for his help. I am also grateful to the hard work and diligence of Anthony "Corky" Carter, Gail Snyder, Joe Dagen, Susan McCauley, and Bruce Brazas while reviewing my work.

Dave Johnson's insightful suggestions helped me shape the direction of my ideas, and help direct the book toward wider applicability.

Scott Geller's thoughtful comments about *Building Safety Commitment* were also extremely helpful and valuable.

The Association for Contextual Behavioral Sciences (ACBS) community, an incredible group of scientists and practitioners, is the major resource for the concepts in this book. Because of the brilliance and generosity of this group of human behavior experts, I have learned the foundation for helping people maintain their commitments. I am extremely thankful for all the great colleagues I have in the ACBS, especially the folks who directly inspired this book: Steven C. Hayes, Kelly G. Wilson, and Frank Bond.

Ron Derby created graphics that fit my vision for the book and helped me visually express my ideas, and I am grateful for his help. Thank you so much.

Thanks to all of my friends for their ongoing support. In the movie *It's a Wonderful Life*, George Bailey says, "No man is a failure who has friends," and since you are all in my life, I feel like the most successful man in the world.

Finally, I would like to thank my family for their patience with me whenever I disappeared to my office to bang on the keyboard to get this book done. Thank you Harmony and Louden. Someday I hope you'll thank me for lecturing you all the time about being safe, committed, and mindful.

Words can't express my gratitude to you, Jen. Our commitment to each other gives me safe passage in this life. Thanks for being everything to me.

Introduction

My dad was an electrician who helped build the World Trade Center in New York City back in the early seventies. Even though I was very young, I can remember him coming through the door after work and dropping his yellow hard hat, clunky goggles, and spools of colored electrical tape in the front hallway. I can also recall when we were playing together, I'd put his hard hat on my head and run around the apartment in my diapers.

My dad was also a safety fanatic and he instilled in me that same strong value of safety. So much so, that the people in my family call me "Dr. OSHA" because of my dedication to childproofing their houses. If you ever have a baby in your home, and need plastic covers for your electrical outlets or cushions for your coffee table corners, I'm your go-to guy!

Today, I work as a safety consultant, and I've been fortunate to work with many excellent companies in various industries. I am the Senior Vice-President of Quality Safety Edge (QSE), a recognized leader in the use of effective methods for improving safety and performance in the workplace. In the past, I worked as an electrician in New York City skyscrapers, a roofer in rural parts of the Midwest, and a safety evaluator in dangerous work environments in many parts of the world.

Ever since I first stepped my steel-toe boots on a construction site in 1988, there's never been a question in my mind that hard hats, proper eyewear, and work gloves are essential to personal safety. But I quickly came to understand that safety has many other important components in addition to wearing personal protective equipment. Focusing on occupational health issues, creating effective safety processes, promoting a safety culture, and having solid leadership involvement, among many other crucial topics, help ensure people will go home happy and healthy after a hard day's work.

After 20 years of studying people's behavior, I wish more people knew about another fact I am absolutely sure of: to avoid injury, you've got to have a personal commitment to safety. To maintain your own safety and the safety of others, you have to know what to do, be motivated to do it, and make sure your "head is in the game." You see, a hard hat might protect your head, but it is what is underneath the hard hat that keeps you safe!

You have been provided with many resources, such as PPE, training, and leadership guidance, to help you deal with the external working world. But your internal-world, which are the thoughts and emotions you bring to the work site, also have a major impact on your actions. What goes on between your ears and behind your eyes is a critical ingredient to your safety.

Safety training does not typically teach you how to deal with your internal-world of thoughts and emotions. It doesn't typically address how to maintain focus or how to handle distracting feelings that arise in the workplace. Safety workshops rarely address how to deal with diversions that can take you mind off your task, and in turn, put your safety in jeopardy. This book will provide you with a method for managing your internal-world so you can handle all that.

Another problem that arises is the temptation to save a little time or energy by taking risks and shortcuts at work when we know better. Sometimes, we know we are breaking a rule and we roll the dice anyway. Our gut tells us the probability is low that we will get hurt, and we succumb to that temptation. We take small risks, not because we are thoughtless or irresponsible employees, but because we have not gone through a process of self-examination that helps us look at our behavior in the light of much broader issues in our lives. When we examine what is meaningful to us, and become aware of our values, it influences our work behavior forever. This book will take you through that process of self-examination. I hope when you finish this book you will agree it has changed your life and your thinking about how you work.

Of course, the objective of this book is for you to be more personally committed to safety after you read it. What is your current level of safety commitment? Maybe you are a safety champion at work and volunteer your weekends to teach Safety Merit Badge class for your local Boy Scout troop. Perhaps you are a bit skeptical about the safety procedures at your workplace, and you don't really see the point of all the precautions. Maybe you recently got injured or witnessed a close call. Or perhaps you fall into the category of folks who just don't care about safety because it slows down production and seems kind of wimpy.

No matter your level of safety commitment, since this book is in your hands right now, you probably work in an environment where there is a risk of getting injured. It doesn't matter if you work in a fairly calm environment where the worst threat comes from bad ergonomics, or if you work somewhere that Mike Rowe and his *Dirty Jobs* television camera crew would refuse to go. Whether you picked up this book yourself, got it from a friend, or were told to read it by your boss, I appreciate your willingness to explore the ideas we'll be talking about, and I believe you'll find it worthwhile. Regardless of how you got here, I hope you look at this book as an opportunity to learn something new and potentially exciting— something from a very different point of view than you're used to hearing about safety.

Some of the topics we'll discuss in this book will be unique, even if you have been working in your industry for decades. We'll talk about how to strengthen your personal safety commitment by building a Safety Commitment Plan. We'll also talk about how everyone gets distracted from time to time, and what can be done to expand your "situational awareness." In addition, we'll be talking about your own internal obstacles to commitments, even going as far to discuss how your thoughts and emotions can become significant hurdles to following through on safe actions. After years of training people to build their commitment skills, I am confident some of the subjects we will cover in this book will be different from any of your previous safety training. And please keep in mind, sometimes it's important that we stretch our boundaries and start to take a new look at things because that's what progress and improvement is all about.

But before we jump into that arena, I want to let you know a little bit about myself.

I used to work as an electrician's apprentice during the summer of 1989 on the construction site of the J.P. Morgan Bank in New York City. My father was part of the International Brotherhood of Electrical Workers, and when the children of union members in Local 3 were in college, they were offered employment in a summer job program. This job was a great opportunity for me to work on the construction site of a famous NYC skyscraper. Since my dad was working somewhere uptown, I was pretty much winging it on this site as an unskilled, lowly gofer.

After I'd been hanging light fixtures for several hours one Monday morning, my foreman sent me to the store to pick up snacks and drinks for the coffee break. One of the journeymen went with me. His name was Hal and he wasn't coming along to help me or supervise me… he just wanted to go outside for a cigarette. But even if his intention was to satisfy his own nicotine craving, on our way down the freight elevator he talked to me in a way I remember to this day. After the elevator doors closed behind us, he grabbed my Local 3 hat off my head. He slapped it across his knee a few times in order to clean off all the dust and debris that accumulated while hanging lights up in the drop ceiling. He said, "You're going outside to the real world. Show everyone you have pride in what you do and how you do it." He put the cap back on my head.

Once we reached the bottom floor and left the elevator, I took a few steps and then he grabbed my shoulder to stop me. He didn't say anything after I turned around to look at him, but his face told me he had something important to say. He fumbled for his cigarettes, put one in his mouth and then started patting around his pockets to find his lighter. I remember being a little impatient because our foreman didn't like it when the other apprentices took too long bringing back the coffee.

Once Hal lit up his cigarette and took a drag, he looked down at the floor and then back up at me with one eye closed. He exhaled all the smoke and at the end of his breath he said, "College, huh?"

I kind of braced myself because I know some of the guys were suspicious of teenagers like me coming to work for three months and then going back to school. This unique work situation seemed to create a rivalry between the college kids and the working men. But before I said anything, he said, "Good. I'm glad you're here because I can tell you honor the work we do. It's the same work your father does. You know what it's like to work hard."

He took another drag off his cigarette and then said, "But I want you to do me a favor." (In New York, that's a polite expression people use when they are going to *tell* you what to do.)

He finished: "Do me this favor. If you learn to do something in college, whatever it is: *commit yourself to doing honest work.* When I put in a light switch, when it's done, it works. When we install a power panel, we know we did honest work because when we're done, it works. Whatever you do, commit yourself to doing honest work. Make sure what you do for people is actually worthwhile."

"OK," I promised.

And then he reached in his pocket and grabbed a five-dollar bill. He said, "Light and sweet," to indicate his coffee order and told me to get something for myself with the change. He held the money tight as I tried to take it, and said one more time, "Commit yourself to doing honest work!" and I knew he wasn't going to let go until I said, "OK," so I said it right away.

I took that promise seriously as a 19-years-old, and I will not go back on it. I am committed to doing honest work, and I will only write about the things I know will improve safety. Some of the things you read in this book might be new to you, but they will be in the service of honoring my promise to do honest work for people I respect and because I have valued safety my whole life.

Chapter 1

The External & Internal-World of Safety

Every single day workers just like you and I get hurt or die at the workplace. Assistant Secretary of Labor for OSHA, Dr. David Michaels said, "Forty years of common-sense standards and strong enforcement, training, outreach and compliance assistance have saved thousands of lives and prevented countless injuries... In 1970, 38 workers were killed on the job every day in America; now it's thirteen a day. This is a great improvement, but it's still thirteen too many." [1]

Safety initiatives have made a substantial impact on preserving life and limb, but if 13 people die on the job every day, we have a long way to go. Governing bodies, companies, and workers have collaborated to help the *external-world* of the work environment become less hazardous. This book will continue in that tradition of making workers safer by helping the *internal-world* of the employee become more committed to safety. By combining efforts to improve the internal and external-world of the worker, we will surely see a continuous reduction in injuries over time.

As one of my favorite safety signs says, "Teamwork makes the dream work!" The purpose of this book is to help you develop your own Safety Commitment Plan which bridges the external and internal-world, and the purpose of this chapter is to take a look at what is meant by "external-world" and "internal-world."

The External-World

A company's first line of defense for protecting its workers is to engineer safety into the workplace by *changing the external-world*. In other words, the external work environment must be kept free from hazards, and people should be protected from potential incidents during the workday by ensuring the environment is safely maintained. A safe workplace should be a company's chief concern.

There are many ways companies and employees contribute to reducing risk in the external-world, from following appropriate safety procedures, to finding and fixing hazards, to wearing PPE, just to name a few. Many work sites are naturally risky due to dangerous materials, exposure to the elements, and constant hustling to meet or beat deadlines. By altering the work site to reduce exposure to hazardous chemicals, tripping hazards, falling objects, and so on, we are changing the external-world to preserve the well-being of the worker.

When you observe the external-world of a work site, you often see evidence of efforts to protect employees. You may see safety *signs*, black-and-yellow striped markers, and reflective symbols everywhere. When I walk into a work site, whether it is a foundry, quarry, mineshaft, factory, power plant, or mill, I see signs all over the place reminding me about the importance of safety and how I should behave. In addition to putting up signs, most companies require or provide personal protective equipment (*PPE*) in order to do a job. There are also requirements related to the *proper tools* and materials to be used on-site. No responsible company gives out grinders without safety guards or power tools without grounded plugs. Safe tools and PPE are part of the everyday work gear for frontline workers. You can see an organization cares about safety by how they set up the external-world to protect the worker.

Companies also provide education and *training* to make sure you know how to act safely. Organizations often provide training in basic safety skills, and the more complex your job becomes, the more complex your safety skills need to be. Veteran workers possess a great deal of education and training, not only about their craft, but also about maintaining safety.

In addition, industries have also worked to improve *leadership* and management practices to make sure the work is guided toward safety. Leadership involvement has a significant impact on the safety participation of front-line employees. There are many leadership training initiatives, management systems, and culture building programs that can improve how organizations operate. Over the years, we have learned better ways to lead people to be efficient with their work practices without sacrificing their security or productivity.

The external-world is also governed by Environment, Health, and Safety guidelines. Standard operating procedures inform leaders, supervisors, and front-line workers about a multitude of work issues aimed to keep people safe. *EHS guidelines* can include something as simple as requiring pylons to be placed in certain areas, all the way to the complex requirements for facility design and operation.

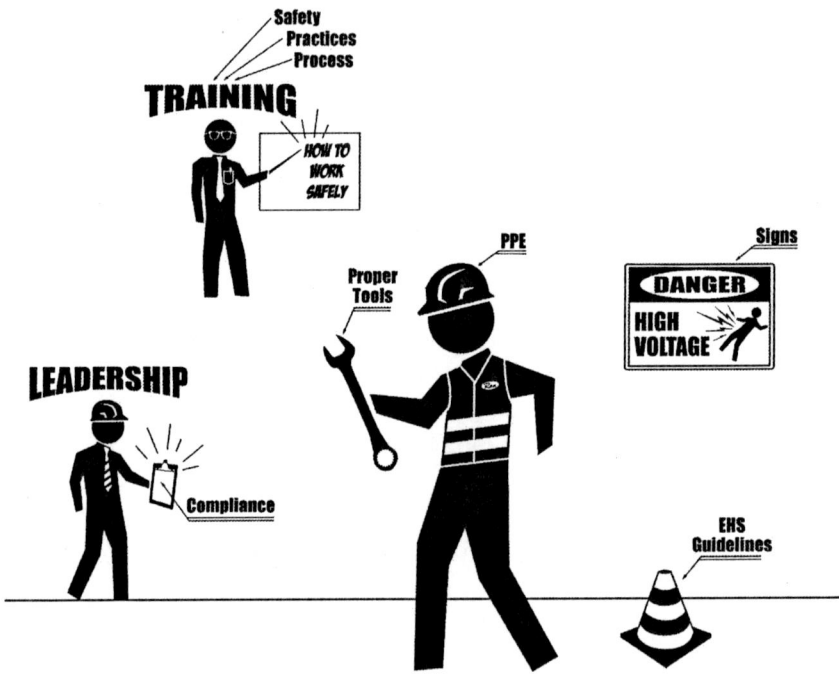

Companies must strive to create sustainable safety in the external-world by using a holistic systems approach incorporating modern technology, engineering, and scientific applications to improve the work environment. And there is also another aspect to safety that needs to be addressed in order to improve safety in the workplace. We need to take another step to eliminate injuries by addressing the *internal-world*. Before discussing the internal-world, I'd like to talk a little bit more about how the external-world helps influence people to behave safely.

Compliance – Motivation for safety from the external-world

Workers receive training about what they should wear, how to use their tools, and the safest ways to act in the external-world. But what gets trained is not always what gets done. Sometimes, we are simply counting on the worker's inclination to *comply* with the minimal standards. Surely, management should utilize a holistic, systematic approach to remove obstacles for executing safety trainings. And it is worth noting that during safety courses, trainers rarely teach workers *how to make a commitment to follow through on all of their safety training.* Rather, the hope is that workers will keep their actions in *compliance* with expectations.

Compliance means acting a certain way because someone else, such as a boss or OSHA, demands it. When you adhere to a certain standard because you want to avoid getting in trouble (fined, fired, or yelled at), you are *complying* with rules set up by your company, industry standards, or government regulations. In some cases, you may comply with the rules for external consequences such as approval of peers, recognition from supervisors, or other positive outcomes. In general, there's nothing wrong with obeying rules or conforming to expectations set up by someone else, but people rarely go above and beyond the call of duty when they are only expected to stay in compliance. There is also a risk that safety compliance will slip when people think they can get away with cutting corners because other people aren't observing their work.

I'm not saying safety compliance programs are bad. Following directions created by experienced workers, and being observed by supervisors in the external-world is an essential part of getting along at work, especially when starting out in a new workplace. Rules tell us how to behave safely so we don't have to learn about occupational hazards the hard way. For instance, when an oil field worker goes into a pumping facility, he initially wears his flame retardant clothes in order to comply with his boss's commands. The rookie is acting safely on his first day in the oil field because he is complying with the rules. He wouldn't dare come to his first day of work without the mandated PPE because he'd be fired on the spot.

Most newbies are eager to follow rules because they want to keep their job and be looked at as a good worker by their boss. However, compliance can deteriorate because workers learn what they can "get away with" and which work tasks have a low probability of leading to an injury. Ask yourself this question: Are there any rules you follow only because someone else asks you to do so? If you answered yes to that question, then you are complying with certain rules on your jobsite.

While compliance is an important component to workplace safety, it's not enough to keep people safe at work. People who've been trained in safety processes are still getting injured or dying on the job. In the United States, thousands of skilled workers die every year from job-related incidents and millions of people around the world die every year from work-related incidents and illnesses.

Behavior-based safety (BBS) was developed in order to answer the call of improving safety in the workplace. BBS is a powerful approach to increasing safe actions at the work site, and often starts out relying on safety compliance, but ultimately aims to help workers become more engaged in the safety procedures. Effective BBS processes lead to employee empowerment. In BBS, all levels of the organization are observed during work and given immediate feedback about whether or not they met certain objectives. Many BBS processes include an incentive or reward to help motivate workers to meet certain safety goals. This process uses a positive approach to create an external-world environment where employees encourage one another to work safely.

In most BBS initiatives, observations are also used to identify and address both unsafe practices and hazards in the workplace. Research has demonstrated the effectiveness of such an approach for reducing incidents and injuries in a wide variety of work settings.[2] The BBS process integrates the work community as an important external-world component for improving safety.

While behavior plays a role in workplace safety, it's important to understand fatalities and injuries can certainly be due to occupational health & safety problems or environmental events that are unpredictable or difficult to control. Governing bodies, companies, and individual workers have teamed up to reduce risks in the external-world by changing the physical environment, improving hygiene, and streamlining behavioral processes. Yet some skilled people just like you and your coworkers suffer because the advances in equipment, training, supervision, and other external efforts are simply not enough.

What if safety can be improved, not only by technological advances in the external-world, but through new advances for changing your own *internal-world*? You are probably wondering what I mean by that.

The Internal-World

Earlier in this chapter I pointed out how companies put signs all over a work site to remind you to act a certain way while on the job. A majority of these signs alert us of hazards in the external-world:

These types of cautionary signs are important because they alert us of potential external risks. However, you also see other signs at a work site that are not about the external environment. Some signs aim to help you change your *internal-world* of work. They have to do with improving your *personal motivations* for working smart, and improving *how you think and feel* about safety. You will likely recognize some of these slogans from signs you've seen around the workplace:

And we can't leave out the sign that is on almost every job site:

When you read these popular slogans, you can see they target personal issues such as choice, intention, responsibility, and thinking. How we choose to act, what we intend to accomplish, and the things we think about during the workday all contribute to whether or not we act safely. The slogans on these signs aim to alter safety from the inside out. Safety officers hang these signs around the external-world because what you think about in the internal-world has an impact on your safety.

Focused, motivated, and properly trained workers have a greater advantage over distracted, complacent, and unaware workers. The governing bodies, caring companies, and safety-minded workers use these slogans because they see the usefulness of changing how we think and feel about safety throughout the day. We all know there is value in making sure each worker's internal-world (how they think and feel) is focused on safety.

When I talk about the internal-world, I'm referring to the things going on inside us that can influence us to act a certain way. We show up to work with our thoughts and feelings, moods and emotions, and other personal stuff that affects us. Sometimes we clock into work with other things on our mind the job. We might have the Monday Morning Blues or be thinking "T.G.I.F." instead of focusing on our tasks. And if you doubt our internal-world affects productivity at work, then I want you to imagine what the workplace would be like if some people did not drink their morning coffee!

We all have an internal voice that constantly speaks to us. Many of the things we say to ourselves have an influence on how we act, so it makes sense to learn to think more clearly, maintain better focus, and stay strongly motivated to act safely in the workplace. *Building Safety Commitment* focuses on these very important aspects of safety, because our modern work sites are too dynamic (and in some cases, too dangerous) to have safety left up to external-world controls and rule compliance. If we rely only on constant supervision and the rules of others, too much wiggle room remains in the safety process. In many ways, safety is in our own hands. We can be our own greatest protector. This book is about considering what we can do in our own *internal-world* to make that happen.

The Problem and a Contribution to the Solution

In the pursuit of safety, we often look at changing the external-world… and for good reason. The physical world is where the physical dangers are. If there are glaring dangers in the workplace, they require immediate attention. Engineering safety into the workplace and addressing hazards in a factory or a mineshaft deserve our primary focus. Organizations have an obligation to address hazardous conditions and hygiene with constant care.

Let's stop here for a moment, and take the time to answer an important question. Do you agree with that last paragraph you just read regarding safety and the external-world? Take some time to reread it, and see if you agree with it. I don't think anyone would debate the importance of keeping the external-world free from hazards. But I think some people might be skeptical about the need to make sure our personal internal-world is aligned with safety. So here's the challenge: if you believe safety in the external-world is *all that matters*, and you essentially agree with that last paragraph, I want you to reread it, and look at the final word in each of the last three sentences in that paragraph.

Attention. Focus. Care. These three words also describe critical elements of safety, and they are part of our internal-world. To keep a safe external-world, we need to prepare our internal-world for safety, as well. Safety requires our *attention* and *focus*, and we are better motivated to act safely when we *care* about the outcomes of acting safely. This book is about improving attention, focus, and care. We have known for centuries the outside world of work presents many dangers, and we are just now beginning to explore how our personal internal-world can be changed in order to help us preserve our safety in the face of these dangers.

Please understand, lacking safety commitment will not cause an injury, but increasing safety commitment can help reduce the risk of injury. Internal-world challenges, such as distracting thoughts or emotions are never the root cause of an incident. Injuries and incidents occur when there is a combination of factors. A worker is not to blame for his or her injury.

Risk factors in the external-world must be addressed as a priority. Companies have a responsibility to assure safe and healthful working conditions for employees. Leadership must implement comprehensive health and safety programs that find and eliminate unsafe work conditions. This book aims to help workers keep their own commitments to safety and address internal-world obstacles in order to contribute to overall safety, which can contribute to the aim of finding and eliminating unsafe work conditions.

Altering the external-world does not have to be our only line of defense. Altering your internal-world toward safety also helps you defend against potential incidents and injuries. This book aims to bridge the external and internal-world by giving you skills to follow an effective personal Safety Commitment Plan. So, let's take a look at what it means to develop a greater personal commitment to safety.

Chapter 2

Values, Compliance, Commitments,
...and a '66 Chevelle

At the end of the last chapter, I said I'd help you build a greater individual commitment to safety. In this chapter, I'd like to explain the difference between safety compliance and safety commitment, while highlighting the importance of your own personal values for developing a stronger safety commitment. I can clarify my point by telling you a story about a time I went cruising in my buddy's muscle car.

My buddy Joe has a mint condition '66 Chevelle. The engine is supercharged, the interior is clean, and the paint job is a deep maroon color that looks like the last few seconds of a sunset. But after I took one ride in that car, I decided it would be my last one. Let me give you some background to explain why.

Joe and I got our driver's licenses when we turned 17-years-old. For the first year of our driving careers, Joe and I were not the safest drivers on the road. In fact, for the first year of my driving career, I never wore a seat belt. Even when I was a passenger in Joe's beat-up cars, I didn't buckle up. It really didn't seem necessary to me at the time. I had never experienced a car crash in my life and I thought that getting into a wreck was unlikely. So for the first full year of driving, I didn't wear a seat belt.

Shortly after I turned 18 years-old, New York State passed a "Click It or Ticket" law. This new rule stated that if a police officer saw you driving without a seat belt, you'd be given a $50 fine. I did not have $50 to spare when I was a teenager, so this was a very effective threat for me. I started wearing my seat belt more often… or at least whenever I was in a place where I thought I might get caught. So when I drove around busy areas, especially in the daytime, I would "click it" to avoid the ticket. Even as a passenger in Joe's old clunkers, I'd put my seat belt on so neither of us would get in trouble. I was doing this because of *safety compliance*.

The external-world—New York state law—was forcing me to wear the seat belt. I personally was not internally motivated to do so. Essentially I was an 18-year-old kid complying with the law. I was not interested in protecting myself. I was only protecting my wallet from losing 50 bucks. Like many teenagers, I was too young to actually think about and value my own personal safety. (Can you see why safety compliance programs are sometimes a good thing?)

Most company policies aim to do the same thing as state laws: they depend on rules, policies, and codes to influence safer behaviors, and it usually leads to an adequate level of safety compliance. But in these circumstances, most of the employees are working safely simply to avoid a negative consequence and stay out of trouble. I've talked to many workers who wear their PPE because of company fear tactics: "I don't want to lose my job, so I make sure

I wear my goggles when the boss is around." That statement is the core of safety compliance: following a rule so people around you don't deliver a negative consequence. I'm not saying this is a bad approach to safety, but I am saying if a company is only using a compliance-based approach, their safety culture is not as strong as it would be if they could create an environment where each employee learned how to strengthen their own personal safety commitment.

As you realize, one of the major problems with safety compliance programs is when you are unlikely to get "caught" acting unsafely, then you might go ahead and take small risks. Joe and I were unlikely to wear a seat belt when we believed there was a low probability of getting caught. If no one is watching during the workday, you might try to get away with cutting corners on a safety procedure, too. While it is true the negative consequences for non-compliance can influence safety, these negative consequences are not always effective because an authority figure might not be around to dole them out.

When unsupervised, workers who have a weak personal investment in their own safety might not maintain important work standards. In addition, when compliance-based consequences are the only method for promoting safety, workers do not have motivation to go above and beyond to improve safety. As we move forward in this chapter, the important elements for moving beyond compliance toward safety commitment will be highlighted. Let's return to my erratic history with seat belts.

After I spent another few years in New York, and complied with the seat belt laws the whole time, I moved to the Midwest to raise my family, find a job, and start a business. I grew roots in the suburbs of Chicago with my wife and two children, and my career started to blossom. Life was good. However, one day I started feeling nostalgic for New York City. I don't know what I missed about New York. Maybe I was yearning for traffic jams, inflated prices, or a slice of my favorite kind of pizza. Or perhaps I had grown tired of being treated so kindly by Midwestern folks, and simply thought a visit to "the old neighborhood" would help me recalibrate my healthy level of self-esteem. I decided to spend a long weekend in New York, so I called my old friend Joe to let him know I was coming home. He told me not to rent a car for my New York vacation because he wanted to be my chauffeur for the weekend. He said he had a "surprise" for me.

When I walked out of the airport exit, I saw Joe and his "surprise." He was sitting in his beautiful 1966 Chevelle. He jumped out of the classic automobile to say hello and help me with my bags. I walked around his car admiring his prized possession, and then he invited me into the machine. We were going cruising and I was excited to be riding shotgun. As I slid into the car, Joe put the car in gear and we rolled out of the passenger pickup area.

I reached for my seat belt and I couldn't find it on the first try. The buckle wasn't where I thought it should be. I started glancing around, looking over my shoulder, digging under the seat and behind my back to find the seat belts. I looked at Joe and he was already laughing at me. He said, "This car was built in 1966. There are no seat belts in a '66 Chevelle, man!"

I was uneasy, which is not the way I wanted to start my vacation. I had ridden shotgun with Joe for hundreds of hours of my life without a single fender-bender. I personally had driven around the streets of New York for thousands of hours as a teenager and I never wore a seat belt back in those days. In addition, I still had never been in a car crash in my life. But there I was, looking at the dashboard of this classic car, feeling a sense of dread because I couldn't buckle up. And I was asking myself, "Why am I so motivated to wear a seat belt?"

I remember telling Joe I was uncomfortable without a seat belt. He called me a few names that I probably shouldn't repeat, and then said, "We won't get a ticket! This car is above the law. And even if we got one, I promise I'll pay it!"

With that phrase, Joe took away any safety compliance issues. Authority figures had no negative consequences over us. And even if we did get in trouble, Joe would pay the fine for me. But I didn't feel much better about the situation. *Can you think of why I was still motivated to wear a seat belt?*

If you asked me that question, I would tell you I had my own personal motivations to act safely. I would tell you about what I value in this world, and that those things have a great impact on my actions. I have a family I love and an occupation I enjoy, and both of those things act as major influences on my motivation to buckle up. I have people and things I care about that make me want to protect my health and well-being. The purpose of my safe actions was personal and very clear to me: I care about my family and my career. I wanted to be safe for my own reasons. *I was acting as my own authority.* In other words, the police didn't have to look out for me because I was policing myself. I didn't need to have the negative consequence of a ticket hanging over my head. Wearing a seat belt was not about sidestepping a fine, but stepping toward a life guided by my *values*. I was willing to make a personal commitment to safety and I was motivated by the things that are important in my life.

Riding in the '66 Chevelle put me in a situation where I had to make a difficult choice. Joe wanted to chauffeur me around all weekend in a car where I could not act in a way that was consistent with what I care about. Since I really care about providing for my children and being with my family, I couldn't continue to ride around New York in a vehicle that might jeopardize my ability to do that. Because the job I love doing is also linked to my well-being, riding in a car that didn't protect me from injury was not an appropriate choice for me. My personal values linked my committed actions to what I care about. Safety had become personal because I was clear about what was important in my life. Riding around in a muscle car all weekend didn't mean a darn thing compared to protecting what I care about. I told that to Joe and he understood.

I asked Joe if he could recommend a rental car company in the area that would loan me something that would keep me safe while I drove around the streets of New York. He joked with me that rental agencies don't loan people World War II tanks, so we settled on something with airbags and seat belts. I was much happier driving that rental car over the weekend because I had followed through on my commitment to my values.

Values Strengthen Committed Actions

Knowing what you value in life can strengthen your ability to commit to important actions, and also helps you make tough choices in life. Telling my old friend I didn't feel comfortable in his prized possession was not easy. Caring about my safety is a personal value of mine, and I choose to let *what I care about* guide my actions. You can do the same thing with your own workplace safety. You can choose to let what you truly care about guide your actions.

As you can imagine, I was embarrassed to tell Joe I didn't feel safe in his car, but do you really think I should have simply kept my mouth shut when I knew the potential risk of injury? I wonder how many people get hurt every year because they are simply *afraid* to say something about risks. Sometimes doing the right thing can make you feel uncomfortable, embarrassed, or require extra effort. But when you value safety, you don't let your feelings and emotions stop you from acting safely. In other words, you act in the direction of what you care about even in the presence of obstacles. You choose to follow your own personal values without letting your emotions dictate what you do. So let's take a closer look at values because they are at the heart of keeping a commitment.

Defining Values

Talking about values can become complicated, but I want to keep the conversation simple. Values are the ideas you believe to be important, and these ideas help guide you toward doing what you care about. When people talk about their values, they typically speak about important ideas such as honesty, family, health, making a difference, inventiveness, leadership, liberty, having fun, integrity, and many other ideas such as these. Values are the key principles we use to help us link our actions to a sense of purpose in our lives. Life is full of choices, and values help guide our choices toward more personally meaningful outcomes.

In Chapter 4, I will guide you through a series of exercises to help you begin to clarify your values. These exercises will be important because people who are clear about what they care about in life tend to find themselves more committed to the choices they make. A well-established foundation of values is important for formulating a Safety Commitment Plan. Please understand I do not plan to tell you what you should value in your life. My job is not to instruct you about what is important.

Many different things have influenced your personal value system: your family, friends, culture, religion, educators and so much more. The company you work for will likely have an influence on your values. Some people have strong family values while other people choose to make their life about being independent. Some cultures influence people to achieve goals independently, and other cultures inspire people to work collectively. Real problems can arise when your values are not clear to you. Without being clear on what you care about, you might begin to act in an aimless, unproductive, and even risky manner. Once you are clear on what you care about, you can link those important things to strengthening your actions and commitment toward safety. Let's take a closer look at how values work to help motivate you toward effective action.

Exploring Personal Values

We all get to live one life on this planet. And our life is made up of the choices we make and what we do. We may not get to choose all the things that happen to us, but we can choose how to react to those events. It's like you are the captain of your own ship: you may not be able to control the wind, but you can control the sails. Your values can help you chart your course and give direction to your actions. In other words, certain events will happen in your life that you cannot change, but you do have some influence on your reactions to those events. Knowing what you care about in life helps you react in a personally meaningful way.

When you are clear about your values and all the things that are important to you in your life, you can focus your energy and attention toward attaining the goals related to your values. Let's use an example outside of the safety industry. Suppose you truly value the idea of being a good friend. If that is true for you, then you are more likely to call the people you care about to ask them how they are doing, spend time hanging out with them, and sacrificing some of your time and resources in order to help them when they need it. Each one of those actions requires effort and focus.

Now suppose you were planning to sleep late on a Saturday morning, and your telephone wakes you up at 7:30 a.m. You let the answering machine pick it up so you can just stay in bed. But your friend's voice is on the line and you hear her say, "I'm really stuck. My car broke down while I was at the supermarket getting milk and eggs for my family's breakfast, and now I have no way to get back home to them. Can you help me out?"

What do you do? No one is watching whether or not you help your friend. There is no rule saying you *have to* pick her up. There is no Friendship Compliance Program in place making sure you follow through on her request. In fact, the only thing in place at that moment is your tired body in your comfortable bed! All the good stuff you desire is right there in your bedroom with you. Prior to that phone call, you were doing exactly what you wanted to do. You can have *immediate gratification* of all your wants and desires for that particular moment if you just roll over and go back to sleep.

But you value friendship, so you kick off the covers, pick up the phone to tell your friend you'll come get her, and hop in the car early in the morning because you are committed to doing the right thing for the people you care about. You don't *have to* do this, but rather you *choose* to do this. If you didn't value friendship, you might continue with the immediate gratification of your needs by staying in bed. Knowing what you value is a strong motivator for action, even when it is difficult or annoying. If you always decided to do the things that led to your own immediate gratification of pleasure and avoiding frustration, then you probably wouldn't make or keep very many friendships.

Going the extra mile to be a "good friend" sometimes requires giving up immediate gratification, and focusing your energy and attention on something that might not have immediate payoff. Acting like a good friend can lead to the long-term payoff of having people around you in the good times and the bad times. It will add quality to your life. Does waking up early on your day off add value to your life? Perhaps not in the short term, but being there for your friends in their time of need will probably benefit you in the long term. And think about this: does sleeping late instead of rescuing your friend add value to your life? Perhaps it does in the short term, but not in the long term. Now let's return to discussing safety.

Exploring Safety Values

Knowing what you value can be a strong motivator for safety, too. If you aren't clear about what you value, then you are prone to make the easy choices that lead to immediate gratification. You will choose things that lead to immediately *getting pleasure* and *avoiding difficulty* while on the job. If that is how you always base your choices, it will lead to long-term problems with safety.

To continue along the same lines as the friendship analogy, suppose you were planning to work safely on a Wednesday morning at the jobsite, and then your peer comes to talk to you when you punch in at 7:30 a.m. You hear her say, "We have contractors in the building today doing very noisy work, so everyone should wear noise protection headphones instead of the standard earplugs." You are comfortable in earplugs you are already wearing. You hate wearing the headphones.

What do you do? No one typically watches you when you do your job. There is no employment rule saying you *have to* put the headphones on. There is no safety compliance program in place at your building making sure you follow through on her suggestion. In fact, the only thing that is in place at that moment is your desire to be comfortable during your workday. Prior to that conversation about headphones, you were wearing exactly what you wanted in your ears. You can have *immediate gratification* of your desires if you just do what you prefer rather than take the extra safety precaution.

But you value your health and safety, so you remove the comfortable plugs and put on the clunky headphones because you are committed to doing the right thing. You don't *have to* do this, but rather you *choose* to do this. If you didn't value your own health, you might continue with the immediate gratification of your needs: you'd keep wearing subpar PPE. Knowing what you value is a strong motivator for action, even when it is difficult or annoying. If you always decided to do the things that led to your own immediate gratification and avoiding frustration, then you probably wouldn't act safely on the job as often as you should.

When we are in the workplace and make decisions based on immediate gratification, we don't always make the best choices for safety. For instance, wearing proper PPE, such as noise protection headphones is not as immediately gratifying as wearing earplugs. Wearing steel-toe boots and a hard hat is not as comfortable as wearing sneakers and a baseball hat. Taking shortcuts on the job is a way of helping us avoid the strain of hard work while obtaining the pleasure of completing a job sooner, but think about the consequences if you constantly chose the immediate gratification of cutting corners and not wearing proper PPE. In an effort to avoid discomfort and get pleasure in the short term, you would be making unsafe choices that can catch up with you in the long run.

But suppose you were clear on what you cared about in your life. Imagine you knew that you truly cared about preserving your health, providing for your family, and earning a good salary so that you could contribute some money to charity. We could say that those values are things you care about, and they can serve as motivations to behave safely. When you realize you risk losing the things that are dear to your heart because you are cutting corners, it suddenly refocuses your motivation to act more safely. In other words, when you are sure of the things that add vitality and meaning to your life, then you are less prone to do anything that would jeopardize those values. You see, so many safety initiatives tend to be about acting safely because you've been *told to* act that that way. A lot would change if you acted safely because you *chose to* act that way based on your own personal motivations.

We can supercharge safety programs by strengthening each employee's commitment to safety. My goal is to help workers choose to work safely based on their own values, and to link these values to a commitment to safety. Safety programs based on consequences from the external-world are effective, but if you can change your internal-world too, you become more committed to the safety process by making it personally valuable.

If you make an effort to use this book to help you learn to think about your safety differently, you are likely to change your internal-world in a way that will help you be more productive and secure in the workplace. By building your own Safety Commitment Plan, and applying the skills you learn in this book, you will improve your own personal commitment to safety. Let's turn our attention over to what we really mean by the word *commitment*.

Chapter 3

Defining *Commitment*

Peter Drucker, one of the most influential experts in the history of industry said, "Unless commitment is made, there are only promises and hopes." This means commitment is more than just lip service. People often *talk about* the importance of safety, but talking about safety isn't enough to make it happen. Along the same lines, everyone *wants* to work in a safe environment, but if *wanting* something were enough to make it happen, almost everyone would have a winning lottery ticket, a slim waistline, and a stress-free lifestyle. We all *believe* safety is important, but if *believing in* something made it happen, we'd all have a healthy diet and no bad habits. Even further, almost all workers are *trained* in how to behave safely, but if *training* were enough to create optimal performance, then everyone who's been through driver's education would be free of speeding tickets and traffic violations. Ensuring safety is not only a matter of talking about it, wanting it, believing in it, or training people how to act safely . . . it is also a matter of *committing* to it. As we go through this chapter, we are going to take a look at what it means to make a "commitment."

Unpacking the Definition of *Commitment*

Committing means *acting in the direction of what is important to you even in the presence of obstacles*. Since the aim of this book is to help you learn how to keep your commitments, let's dissect that definition:

1) "acting"—As you can see in the first word of the definition, committing requires action. Just talking about doing something is not committing. When it comes to safety, many people *say* they are committed to safety outcomes, but they often do not *act* safely. Acting safely is a definite requirement to your safety commitment, and later in this book we will discuss how actions are critical to a Safety Commitment Plan.

2) "in the direction of what is most important to you"—This part of the definition suggests commitments are personal and based on your values. Long-term dedication is maintained when you have a reason to care about the process and outcomes of your effort. Linking your own value system to your safety tasks will increase optimal performance. When acting safely is personally important to you, your commitment will be strengthened, which is why we will spend time exploring your values later in this book. In the previous chapter, I discussed my personal values related to my family and my career, and showed that what I care about helped me preserve my safety. That line of thinking helped me stick to my commitments and it can help you, too.

3) "even in the presence of obstacles"—This last part of the definition reminds us that accomplishing important tasks is not simple or easy. When you set out to achieve a goal, you should devise a plan for dealing with complications, because hurdles will inevitably arise. Committed actions do not stop when difficulties occur, and we will spend several chapters of this book giving you tools to deal with any internal-world obstacles that impede your commitment. Acknowledging there are stumbling blocks in your path, and then working to solve them, is the hallmark of committed actions. To quote the great Vince Lombardi again, "It's not whether you get knocked down, it's whether you get up." The later chapters in this book will address how to get up when your internal-world obstacles knock you down.

The combination of these ideas explains what we mean by making a commitment. Committing means working toward a goal that is meaningful enough to propel you forward even if problems arise, and even if you "don't feel like it." This book will provide the steps for developing a Safety Commitment Plan that will propel you toward acting more safely.

Preparing to Make a Commitment: The Safety Declaration

Making a commitment requires some preparation. Earlier in this chapter, I said some people only talk about safety without committing to safety. Talk is cheap, and wanting safety and believing it is important isn't too expensive either. You need to actually work at it. But having said that, you still do need to make the decision that you really care about safety. Before you can engage in the actions required for a commitment, you do need to articulate that safety is important to you. When you are preparing to make a personal commitment to safety, you might make the following Safety Declaration:

> "Because my health and the health of others is important to me, I am willing to address external and internal-world obstacles that jeopardize safety, and to work in a manner that reflects safety as my top value."

That Safety Declaration seems to resonate with almost everyone I've ever worked with. Hopefully, you agree with it and it feels right for you. (If not, we can address your concern later.) Making that Safety Declaration is a way to prepare yourself to improve your personal safety. However, please understand that making this kind of declaration is simply priming you to take a first step. You can't just print out that phrase, stick it to your wall, and expect it to magically keep you committed. Making that kind of clear statement about your commitments is certainly necessary, but it is not sufficient for safety's sake. As I said earlier, talking, wanting, and believing is not enough, but it is a decent start.

Whether you post that phrase on your corkboard, memorize it, or tattoo it on your arm, it's still just a bunch of words. The words can guide your actions, clarify your values, and help motivate you through the hard times, but it is important to understand that your commitments to safety require action, personal motivation, and skills that will help you when your internal-world puts up obstacles. To create a strong commitment to safety, you need to know *why* you will commit to safety, *how* you will deal with challenges that arrive during your personal pursuit of safety, and *what* steps to take in order to behave safely.

The Safety Commitment Model

To help you visualize all the components to making a commitment, we use the six-point Safety Commitment Model throughout the book. Notice that "Safety Commitment" is the focal point for the model and there are six domains that contribute to Safety Commitment. The following six chapters are dedicated to each of those domains, and are aimed to help you build your skills in each area to strengthen your committed actions. The six domains are divided into three steps. The next three sections of this book will delve into these steps with more detail, but first, let's take a look at an overview of the model.

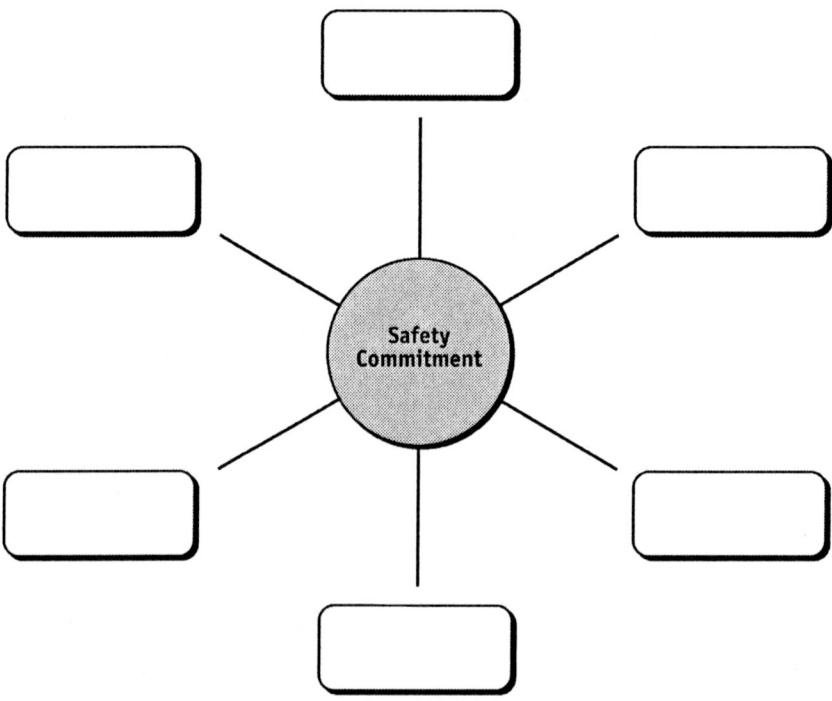

The First Step: Values Clarification

When you take the important steps for improving safety, your performance will be accelerated if you know *why* you are choosing to take on those actions. As mentioned in our earlier definition, commitment requires that your behavior is focused "in the direction of what is most important to you." In other words, a commitment is strengthened when you clarify what is meaningful and vital in your life, and when you use those key values as motivation to behave correctly and safely. Values clarification can motivate and intensify your commitments.

The Safety Declaration in the earlier section states you will aim your actions toward safety "because my health and the health of others are important to me." And since we are talking about values clarification, here is an important question: *Is your health and the well-being of others actually important to you?* If you can answer "Yes," that is great! When you are able to clearly state why you want to engage in a Safety Commitment Plan, then you will likely have the enthusiasm for working through any internal-world obstacles that arise.

But what if your answer to that question is "No"? What if health and well-being aren't things you value very much? Don't worry. Your secret is safe with me.

The health and well-being of others, and even yourself, may not be the foremost motivator in your life. There's no way I'm going to try to change your value system or make you feel guilty for having your own values. There are about seven billion people on this planet and we don't all value the same thing. But we do all value *something!* I couldn't possibly know what that is for you, but I suspect if you and I talked for a little while, we'd be able to clarify some of the things that are vital to you. Maybe you really want to make your life about things such as recreation, money, or being liked by other people. My invitation would be for you to link your safety actions to those things you have made important. In other words, you certainly care about something important in your life, so the challenge is to channel your energy to act safely in order to preserve and enjoy those particular things! You can aim your safety actions to ensure you are healthy enough to do your favorite recreational activity, make more money, and spend time with people who like you.

Later, I'll tell you a story about a group of guys I worked with, and no one believed they had any values at all! After I hung out with them for a while, we were able to uncover they had *unconventional*, but strong personal values which ultimately were successfully linked to improving their participation in safety commitment plans. In the upcoming chapter called ***What I Care About*** we will work together to deepen your own understanding of *why* you care about safety. For now, it will serve as the first component to our six-point Safety Commitment Model.

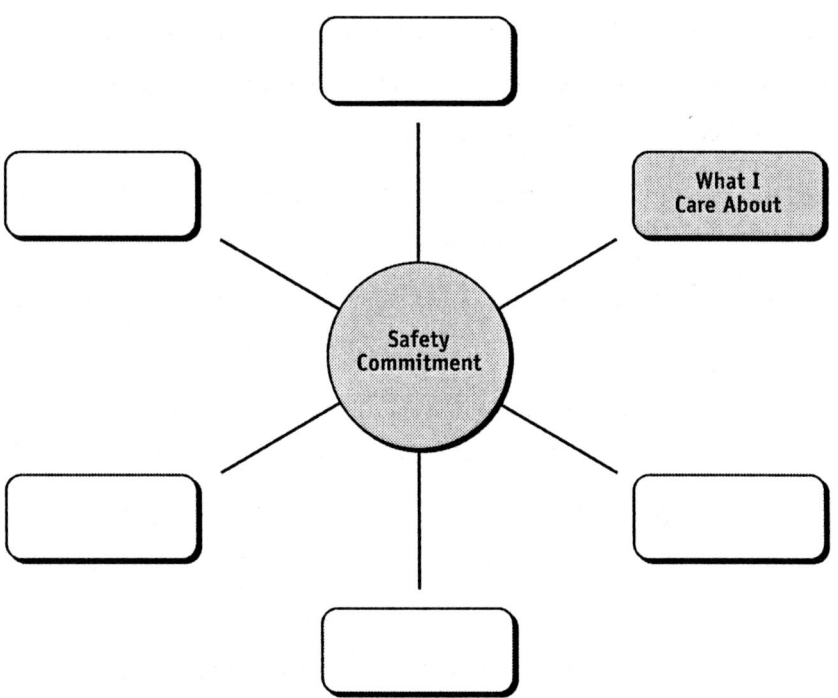

The Second Step: Skillfully Dealing with Internal-World Obstacles
If you recall our earlier definition of *commitment*, you will
remember it requires working toward what you care about "even in
the presence of obstacles." In many situations, hurdles and obstacles
make it difficult to choose to act safely all the time, but if we truly
care about improving safety, these obstacles must be dealt with. I
would like to clarify there are two types of obstacles for safety:
external-world and internal-world obstacles. There are problems you
can detect in the external-world and these need to be removed as best
as possible. For example, if loud machinery noise can be reduced
through better engineering, the people who can fix that problem
should do so as soon as possible. If lack of proper equipment
impedes you from following through on your commitment, make
efforts to acquire the appropriate tools from the people who are

accountable for providing the equipment. If the external-world has safety problems, then workers, supervisors, and leadership should immediately address these problems.

On the other hand, there are sometimes internal-world obstacles to safety, so an important step in strengthening your commitments requires learning how to skillfully deal with these obstacles. For example, sometimes you might not feel like working safely, you could be distracted by things you have on your mind, or you have thoughts that pose a challenge to doing your best on the job. These kinds of internal-world obstacles require a different approach from external-world obstacles because, quite frankly, your feelings and thoughts are really difficult to remove! According to decades of research, trying to alter or eliminate emotions and thoughts is actually *unhelpful*. It is actually *more helpful* to learn how to be safe *in the presence* of those internal-world challenges. You can learn *how* to deal with internal-world obstacles that come up during your daily pursuit of safety.

Four of the upcoming chapters are entitled *Acceptance, Noticing, Here Now*, and *I Am*. In each of these chapters you will acquire four new skills that can be applied to four safety challenges: 1) emotions, 2) thoughts, 3) distractions, and 4) descriptions about yourself. Each chapter will guide you through a set of experiences and exercises to help you deal with each one of those internal-world obstacles to safety. In the meantime, let's add each new idea to our six-point Safety Commitment Model. As you move forward in the book, you will see how each one of these topics contributes to safety commitment.

The Third Step: Safety Commitment Planning

Our definition of *commitment* explicitly states you need to engage in action for it to be a true commitment. As stated in the last section, you need to know *what to do* for safety. The Safety Declaration we used as an example said you will "create and maintain a safe work environment as an ongoing process." To execute that plan, you need to describe *what* behaviors are required to make that happen. Each industry, company, jobsite, and employment position has specific methods for achieving productivity and optimizing safety.

Creating solid, efficient and effective To-Do lists that workers can use as a checklist for ensuring safe execution for their jobs is a simple yet important way for improving safety commitment. This third step of safety commitment helps complete the framework for what you should be **Doing** to improve safety at the work site, and completes our Safety Commitment Model.

Now that you have been introduced to the Safety Commitment Model, please understand the upcoming chapters will focus on teaching skills from each of the six points of this framework. When you bring all six of these points together, you will have learned how to improve your own internal-world experiences and build more powerful action patterns. You will be able to say:

"I am here now, accepting the way I feel,
noticing my thoughts, while *doing what I care about."*

This simple statement is the bedrock for a strong and enduring commitment to safety. Imagine a workforce coming to their job tasks with this kind of committed attitude. Even further, imagine what it would be like if you brought this kind of thinking to your daily tasks. What if, throughout the day, you could keep yourself focused on the task you are doing in the present moment? What if you had the skills to maintain your safety behaviors even when you didn't feel like doing them? What if you could stick to your safety process even when you are having distracting thoughts? What if you could also make sure you were motivated to follow through on your commitment plans because they were personally relevant and meaningful? This book aims to turn "What if...?" into "What is!" Stick with the principles in this book and your behavior will change for the better.

In order to make these principles more practical, we will also work together to help you learn to create your own Safety Commitment Plan, and suggest that workers throughout the company are encouraged to create and follow their own Safety Commitment Plans.

Safety Commitment Plan Worksheet

I am here now, accepting the way I feel,
noticing my thoughts, while doing what I care about.

	Addressing Internal-World Safety Issues	✔
I Am		
Here Now		
Accepting		
Noticing		

	To-Do List for External-World Safety Actions	
Doing		

	Values-Based Motivation	
What I Care About		

Performance Management Contract

Signs and signals are in place

Publicly announced safety commitment

Accountability partner: _____ is aware of my commitments

Describe incentives and performance criteria related to safety process:

Building Safety Commitment Plans

The Safety Commitment Plan is an organized strategy providing specific steps to ensure 1) your mind is focused on your task, 2) your actions are driven by important consequences, and 3) your behavior follows a reliable method for working safely. The Safety Commitment Plan helps you deal with internal obstacles (i.e., emotions, distracting thoughts, and unhelpful self-descriptions), and also maps out how the external-world can be altered to ensure optimal safety while completing a task (i.e., performance management, task checklists). To put it simply, a Safety Commitment Plan combines directions for *addressing internal-world safety issues* with a detailed *to-do list for safety actions*, while encouraging safety with *values-based motivation* and *performance management contracts*. Those components of a commitment plan are discussed throughout the rest of the book. (You can find a blank copy of the Safety Commitment Plan in Appendix A, and you can also download the Safety Commitment Plan Worksheet from www.buildingsafetycommitment.com. It's free and you will be using the worksheet as you move forward with the book.)

The Safety Commitment Plan Worksheet can be used as a simple checklist. The first four rows in the worksheet target internal-world safety issues, such as unhelpful thoughts, distractions, and problematic emotions. When those four rows are complete, you will have a more focused mindset to prepare you to follow an established To-Do list aimed to safely guide you through a job task. Finally, the checklist reminds you of the different reasons you are motivated to act safely on the job. Each one of the six components serves as a building block to your ultimate Safety Commitment Plan. The illustration of the Safety Commitment Plan in this chapter is an empty outline, and it will be filled in as we progress through the book.

According to research,[3,4,5,6,7] there is measurable impact on work behavior when using this approach for improving commitment in an organization. I have also received significantly positive feedback from workers, managers, and CEOs whenever I use this approach to increase employee commitment to safety. We will be using this Safety Commitment Plan Worksheet throughout the book to highlight the Safety Commitment Model. The phrase "I am here now, accepting the way I feel, noticing my thoughts, while doing what I care about" has had a profound impact on safety and safety leadership. In the next few chapters, we will explore each part of this commitment-strengthening phrase. I invite you to connect with the stories that follow, engage yourself in the personal exercises, and see yourself as having the ability to improve your own safety commitment.

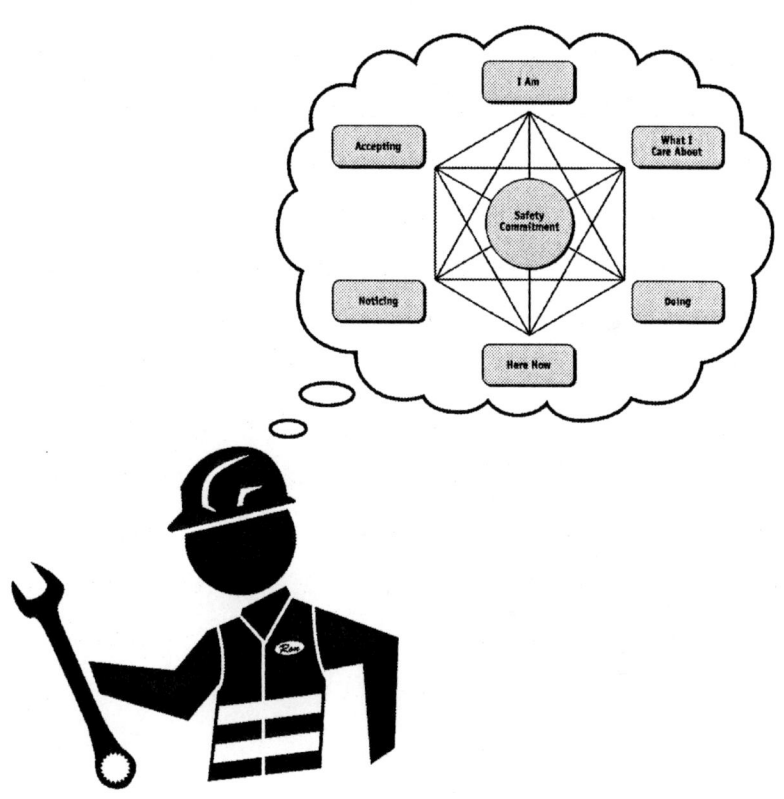

The First Step:

Values Clarification

Chapter 4

What I Care About

Clarifying your values is an important and major step toward true safety commitment. In Chapter 2, I mentioned that values can become a complex subject, but we'll just keep it simple in this book. Values are the key principles you believe to be personally important and help you make choices on how to behave. Roy E. Disney once said: "When your values are clear to you, making decisions becomes easier." In this book, values clarification exercises are used in order to help make safety decisions easier.

Where Do Your Values Come From?

Your values come from lots of different sources: your family, friends, education, religious beliefs, your culture, and much more. What you believe is shaped by your unique life experiences. Since values are personal, you may not have the same values that other people hold dear, and that is because you have had different experiences during your lifetime.

Organizations often try to instill values into members. The U.S. Air Force declares "integrity, service, and excellence" as core values. The airmen who abide by that credo are destined to have a more successful career during their time of service. As an adult, if you value being trustworthy, loyal, helpful, friendly, courteous, kind, obedient, cheerful, thrifty, brave, clean, and reverent, you were probably a good Boy Scout, because those values come directly from the Scout Law.

Values don't just have to be listed as single words either. Harley-Davidson has a company document called *Everyday Values*, and it lists five phrases that depict what is vital to the company: tell the truth, be fair, keep your promises, respect the individual, and encourage intellectual curiosity. As you develop your own understanding of your personal values, you can keep these organizational values in mind as a guide.

Many organizations establish statements based on the company's vision, mission, and values. These declarations help the organization focus on a worthwhile direction and navigate troubled waters. You can do the same thing for yourself to improve your own safety commitment. You can establish your own values statement to guide your efforts toward a worthwhile direction and navigate yourself through times of trouble, too. Having your own values statement can help you keep your commitment to safety. If you need some help, you can investigate your company's values statement and see if it aligns with your values. Or you can do some soul-searching to really illuminate what is personally important to you. We will walk through some exercises in this chapter to help you with values clarification, and before we begin I want to make sure you understand that clarifying your values is an ongoing process and directly related to your safety.

Values clarification is an ongoing process. At different points in our lives, we care about different things. In general, young adults typically focus on different values than adults in the middle of their career. Folks who are close to retirement spend their time caring about different things than early and mid-career adults. This is perfectly natural because what is important to you will gradually change throughout your life. Speaking broadly, it is more likely a young person is focusing on recreational and educational values, while later in life he or she starts to concentrate more on family values and security. It does not have to be that way; I'm simply highlighting an example of how our focus can change throughout our lives.

Also keep in mind that certain foundational values can remain unchanged throughout our lives. It is possible to care about integrity, being trustworthy, and telling the truth throughout your lifespan. When you examine your current values, you may find your attention is focused on the same values as when you were younger, but perhaps not. The purpose of values clarification is to help you know what is vital to you *right now*, no matter what your station in life.

Values clarification is an important process. I am aware many people reading this book will have skepticism toward doing the exercises aimed at discovering the meaningful aspects of their lives. I have personally worked as a front-line employee for many years, so

I know this book is delving into some unusual topics when compared to other types of safety training. But please consider following through on the exercises - even if you are skeptical - because when you are clear on your values, they can act as a powerful motivator for safety.

If you have never really clarified your own values, or if your behavior is not being guided by your values, you might not be living your life in a fully safe and productive manner. Without clear values, you are very susceptible to making your actions simply about avoiding pain and getting pleasure. (Do you remember us talking about that in Chapter 2?) Avoiding pain and getting pleasure is part of our basic animal instinct that helps us survive and many times it works in our favor. But sometimes, in the pursuit of feeling good, people sometimes cut corners. I know plenty of instances where workers tried to finish a job quickly, and because they were rushing, they created an unsafe situation. In addition, sometimes in the pursuit of avoiding uncomfortable feelings, people ignore safety suggestions and regulations. I have been around plenty of dynamic work environments where the workers willfully ignore certain procedures (e.g., lockout/tagout, PPE) because it's a "pain in the neck" to deal with. The basic instinct that guides so much of our behavior – avoiding pain and pursuing pleasure - can certainly become an obstacle to optimum safety. The pursuit of immediate gratification can lead to risking all the things that are important to us.

Think about it. If we only make choices to reduce discomfort and gain comfort, safety will suffer. For example, PPE can be uncomfortable, using tools with bulky guards on them is annoying, and walking through yellow taped off areas is often quicker and easier. The good thing is, most of us wear our uncomfortable hard hat all day at work, use the tool guards despite the difficulty, and take the long way around roped off areas even though it takes longer. Why?

There are two possible reasons we wear uncomfortable PPE, use tool guards, and don't cut corners: 1) compliance with the rules and 2) commitment to our values-based choices. You follow safety procedures because you don't want to get in trouble, or because you care about your safety. (It certainly can be a combination of both reasons.)

Unfortunately, compliance-based safety programs are not always optimal because employees cannot be watched 100% of the workday. When the effectiveness of compliance-based safety is compromised, then the only effective approach for instilling safety is motivation through personal values. When the external-world does not have a good influence on your behavior, you can only rely on your internal-world. When the external-world isn't managing your actions, then safety becomes self-managed. Your personal commitment to safety becomes critical, *and that commitment is fueled by your values*. Knowing what you care about in this world helps motivate you to act correctly. On the other hand, *not* knowing what you care about suggests a lack of personal motivation to act safely. Lacking intrinsic motivation can put you at risk of cutting corners to safety. Hopefully, you can see the importance of values clarification as a motivator for safety.

We will walk through some exercises I hope will put you in touch with your values. I can't promise that when you finish reading this chapter, you'll have all the answers to the meaning of your life, but I do hope that some of the ideas will point you in the direction of what is most vital and meaningful during your time on this planet, and will motivate you to behave more safely on the job.

The Values of Money

I'd like to start your wheels turning about your values by asking you a sincere question. I hope you will be honest when you answer it:

Why do you work?

It is a serious question, so I'll even rephrase it for you:

Why do you do your job?

Whenever I ask these questions in my safety workshops or during one-to-one coaching, I usually get the following answer: "I do it for the money." Sometimes the response is said a little differently, but the answer usually centers on working for a paycheck. I respect that answer and it makes a lot of sense. Most people who work in industries where their safety is sometimes at risk would not do that kind of a job for free. And yet, when I hear money is the reason you are working, I need to ask a follow-up question:

Why do you need the money?

I know that follow-up question seems naïve, but stick with me, because I'm not looking for a superficial answer. I am looking for a personal answer. I know we all need money to survive and thrive. Money buys you food, water, shelter, and pays the bills to help you survive. But money might also help add more to your life than just these basics. It helps you thrive, prosper, and succeed at your goals.

Money helps you become involved in meaningful experiences and acquire things that are important to you. So my follow up questions are critical:

What experiences are you looking for during your life?

What are the special things you want to have?

*Why do you need the money outside
of taking care of your most basic needs?*

Once I start asking these kinds of questions during workshops and coaching, people start to give very different answers. Some people begin to talk about how they care deeply about their family and want to provide for their spouse's and children's needs and desires. Other people talk about paying for comforts of a home, security of health care, and fun experiences on weekends and vacation time. You can spend your money on lots of things, and those things are little hints at the real answers to the question: *Why do you work?*

Beyond the Paycheck

At the end of this section, I'm going to ask you to take a few moments and honestly answer the following question: "Why do you do your job?" I'm going to suggest that you give some thought to your answer so it goes beyond simply acquiring a paycheck. Whenever I ask this question during one-to-one coaching, and after we get beyond superficial answers related to money, we begin to move toward a meaningful values statement. Let's take a look at some broad topics that contribute to why people work. You might agree with some of these reasons for working, and you might not agree with some of the others.

- **A sense of accomplishment** – Employment gives you opportunities to reach a goal or achieve something meaningful. We know that human beings do things all the time just because of the challenge. George Mallory, the famous mountain explorer, said he wanted to climb Mt. Everest "Because it's there." Turning our attention to a different sport, less than 1% of the people who complete marathons actually get paid to run the grueling 26.2 miles. In fact, they actually *pay* entry fees to take part in the race! There are people in the technology world who create "freeware" just because they want to see if they can create a powerful computer tool. Besides a paycheck, your ambition to accomplish great things might be why you work... and keep in mind your pursuit of accomplishment will be hindered after an injury.

- **A sense of contributing to the community** – Another values-based motivation for working is "giving back" to your community. One of my favorite moments as a safety consultant happened during a renovation outage at a humongous power plant in the U.S. I was standing in an inoperable furnace with a boilermaker who said to me: "I may not be able to save lives like a surgeon can, but I save lives by keeping the power on in the surgery room." He was beaming with pride because he was contributing to society. He saw himself as a part of the greater whole, and was motivated to maintain his productivity and safety because he knew his job as a boilermaker helps save lives. Perhaps you work because of how it affects your community. If so, notice how an injury – or worse – could be devastating not only to you, but to the people counting on you!

- **Idle hands are the devil's tools** – Maybe you work because a job provides structure and scheduled activities. Without employment, people often search for things to do with their time, and what they find might not always be good for them. Some research suggests unemployment is a risk factor for alcohol abuse and domestic violence. Keeping yourself occupied and out of trouble can be vital for your lifestyle, and can motivate you to stay safe and on the job.

- **Idle hands are also really boring** – Working provides some sort of endeavor. While many people yearn to just go home and relax after a long week, many people also find that sitting around gets old after a while. In fact, lots of retired people start looking for a job, even when they are financially secure, because they want something to do with their time. Injured people also have a lot of time on their hands. I have interviewed almost a hundred people who were lying in a hospital bed after an injury. An overwhelming majority of them have confided in me that they just want to get up and do *anything*. Many of them would by happy to go to back to the job they used to dislike, just because – let's face it – lying around is a drag.

- **Providing an identity** – Once you've been working at a job for a while, you might continue to do it because it becomes part of how you describe yourself. People derive a great deal of satisfaction by being part of a particular profession or team. Being able to say, "I'm a soldier in Army Rangers Airborne," "I'm with the Carpenter's Union," or "I run a landscaping company" are all statements that help us describe who we are and even know ourselves better. I'm not saying your job should entirely define who you are because sometimes getting too wrapped up in your job identity can be a problem. On the other hand, having a vocation can give you a feeling of pride, belongingness, and purpose. The question is, why would anyone risk something so dear to their heart by cutting corners on the job?

- **Serving a higher power** – I've worked with folks through the years who have confided in me that they work because it is part of their spirituality. My question, "Why do you work?" has twice been answered with biblical scriptures. One worker said, "Those who work their land will have abundant food, but those who chase fantasies have no sense," (Proverbs 12:11) and the other worker said, "There is nothing better for a person than to enjoy their work, because that is their lot in the Lord's blessing" (Ecclesiastes 3:22). Being able to tithe for spiritual reasons, or simply give money to charities can also be a reason for working. I'm not telling you these *should be* your reasons for working, but simply suggesting that some people are motivated to work because of their religious values, and it is hard to do that if you are injured.

- **Work is fun** – Many people outside the industrial professions think blue-collar work is strenuous and demanding. That can surely be an appropriate description, but the work can also be a great deal of fun. I have witnessed the joy on people's faces when "laboring" on a demolition site. One demolition specialist told me about the joys of using a plasma torch to remove metal tubes from a boiler in a paper mill. He said: "This tool is like a light sabre from Star Wars. I really enjoy my job." If your job is fun, you will cease to enjoy it if you get severely injured.

- **Social interaction** – When you spend 40 hours a week at a worksite, you can really get to know your fellow workers. Regular contact helps you become more familiar with each other, and shared experiences help bring people closer. You are very likely to have a few things in common with coworkers, and maybe even set up times to socialize during lunch or after work. Deep friendships can be made, especially after working with a partner or a crew for a number of years. Being able to see these friends regularly not only makes you want to be safe, but you might even be motivated to keep them safe, as well.

This list is merely a few broad strokes to some of the added benefits that come from working. It is not an exhaustive list, and I encourage you to think about expanding the list with your own personal reasons for working. But if you are struggling to think about why you do your job, I hope this got the wheels turning. So I ask you:

Why do you do your job?

_____

_____

_____

_____

_____

What Do You Want to Be Known For?

Thomas Jefferson was a man of spectacular achievement, and he did something very interesting in his old age: he composed his own epitaph. In his memoirs, he described what he wanted his tombstone to look like and wrote that it should have "the following inscription and not a word more:

Here was buried

Thomas Jefferson

Author of the Declaration of American Independence

of the Statute of Virginia for religious freedom

Father of the University of Virginia

…because by these as testimonials that I have lived, I wish most to be remembered."

Epitaph exercise

Inspired by Jefferson's foresight, I pose a similar question to you: For what do you wish most to be remembered? In other words, what do you want your tombstone to say? This is a fairly common exercise people use to help clarify their values. I invite you to grab a pen and truly fill in the blanks for this exercise. See if you can honestly declare how you want your life to be summarized. What are the things that are vital to you? What do you want to be known for? Like Jefferson, can you articulate what is important to you? You don't have to just pick three big accomplishments like he did. Maybe you want to list your relationships, religious affiliation, or how much you love your favorite sports team. You can include qualities of your character such as being patriotic, dependable, or generous. It's your epitaph – feel free to write it in a way that really reflects who you are and what you care about.

Eulogy exercise

Some folks find the epitaph exercise too restricting. If you prefer to expand what you want to be remembered for beyond the epitaph, start writing your eulogy. A eulogy is a speech someone makes about you during your funeral. The speech usually covers the important characteristics and accomplishments of the deceased, and can cover a large number of topics, such as family, work, recreation, education, friendships, etc. Consider writing your eulogy right now on a separate piece of paper. Can you imagine what it would be like after you die, and what people would say when they are discussing the way you lived your life?

In your eulogy exercise, include the things that are so important to you that they are also obvious to people who are giving the speech. If you need help starting the eulogy, they usually begin with: "We bury a very dear person today. This person lived a full life because s/he was very involved in (fill in the blanks)... This dear person also deeply cared about (fill in the blanks)..." What would you want your eulogizer to fill those blanks with? What is special about your personality, your habits, and your qualities? Consider taking a few minutes to capture what you truly care about during this exercise. You can write your eulogy in the blank pages at the end of the book.

Lifetime achievement award exercise

If you are thinking that the last two exercises are rather morbid, and you don't want to ponder what it will be like after you die, then let's turn the next exercise into a celebration of your life. I often use the Lifetime Achievement Award to help people with values clarification. Grab a pen and fill in the next exercise box with personal qualities that are so important to you that people will recognize them when applauding you for a milestone in your work career. Really let loose on this exercise. What would you like to be recognized for when it comes to being a friend, parent, significant other, or spouse? When it comes to your involvement in the community, what are you hoping will be said? What do you value when it comes to your health, hobbies, and education? How involved are you in your hobbies? How did you behave in the workplace? In this exercise, let yourself take a look at your most desired qualities.

If you won a Lifetime Achievement Award, what personal qualities would you like to be recognized for during the award speech?

I recommend taking some time alone and focusing on the values exercises you just went through because personal values can truly motivate you to stick with your safety commitment. When you know what you care about, it is less likely that you'll risk those things by cutting a corner for immediate gratification.

This Lifetime Achievement Award exercise can be a lot of fun when done in groups, too. In some workshop environments, I give out small computer devices that help me survey the audience members about their values. I can summarize their anonymous responses so participants can see how their peers responded. We then create a list of all the things that motivate people to do their job and do it well. It is really remarkable to watch and listen to people talk about being motivated to work – and work safely – for personal reasons that extend beyond receiving a paycheck.

If you've gotten this far without actually participating in the exercises, then please consider going back and taking at least one of them seriously. If you don't want to write in your book, find a separate piece of paper. People who write down their values are more likely to be conscious of them and have their values guide their actions. If you don't have a pen, then at least try them out as thought exercises. If you are struggling with coming up with answers, then the next section might help you because it gives real life examples of people's motivations for working safely. In the end, please understand that your responses to these three exercises will help you create a unique blueprint of values that are personal to you, and can motivate you to do the things that are meaningful in your life, and hopefully motivate you toward acting more safely. They will also play a role in your safety commitment plan.

Examples of How People Respond to Values Clarification

When you broaden your understanding of what you care about in this world, you become more motivated to commit to safety actions. When you are clear on what is meaningful to you (family, health, friends), and what qualities you value (honesty, caring, providing), then cutting corners for short-term gain and immediate gratification is less attractive. Let's take a look at some of the things that motivate other people to follow through on their safety commitments.

When I was working with a large manufacturing company, the top Health, Safety, and Environment officer said he really wanted to be known for his untiring pursuit of safety. He even said, "I know some guys give me a hard time about my constant reminders to be safe on the job, but I don't care about their opinion of me. I just care about their ability to go home at night." He valued the safety of others so much that he stuck to his commitments, even when other people ridiculed him (i.e., even in the presence of obstacles). Safety was the key principle he believed to be personally meaningful and that motivated him to make choices on how he promoted safety at the worksite.

I imagine the world would have fewer incidents and injuries if more people valued safety the way that HSE officer does, but you can also prioritize different values than he does and still be motivated to work safely. I remember one young woman who worked as a welder telling me her life was dedicated to her children. She said she would be fine if her epitaph just said "Mom." She then explained how her values linked up to her safety commitment: "I do my job safely so I can go home to my kids. When I'm at work, I make sure I'm aware of my surroundings because I want to see my babies. On the job, I follow the rules, whether the boss is around or not because that's how I am assuring myself I'll see those kids tonight and every night. My safety is about them." It was personally meaningful for her to be healthy, present, and involved with her kids, and those topics were key principles helping her make safe choices at the worksite.

People respond to values clarification is various ways. In chapter 3, I mentioned I'd tell you a story about a crew of workers who were rumored – by their managers – to have no values at all, but who eventually came around to improving their safety after values clarification. I met this small group of guys while consulting with a demolition company in the southwestern part of the U.S. Before working with the crew, their supervisor told me: "These guys are like the black hat wearing cowboys of the Old West. They'll complete their tasks, but they don't like you to tell them how to do it. They do a good job following through on the 'Git-R-Done' work ethic, but they don't take the time to focus on safety! I don't like it, but they're the only guys I can find who'll take on this kind of job!"

This company's demolition work was not only dangerous, but it was fairly seasonal, too. Depending on the time of year, there might not be work for six weeks straight. Given the location of the company, the inherent risk in the job, and the lack of a steady paycheck, this supervisor did not have a lot of flexibility when it came to hiring employees. The men willing to do this job had certain characteristics: they liked to live in a wide-open part of the country, were macho and fearless, and didn't mind being unemployed for weeks on end. These same characteristics were probably the reasons why each of the guys in this crew also dedicated his life to a motorcycle club.

The supervisor told me he was having a hard time getting the guys in the motorcycle club to follow compliance-based safety programs. I suggested that instead of telling them to comply with rules because an authority figure said they *had to be safe*, we could invite them to make their own commitments because they *want to be safe*. The supervisor laughed, and when I told him I would do this by helping the crew base their safety actions on their own personal values, he laughed even harder and said, "They ain't got no values." I assured him everyone has values.

During scheduled small-group meetings held at the company headquarters (and admittedly, one meeting was held at a tavern), the workers and I discussed why they did their job. Inevitably, the conversations centered on money, but once the guys trusted me a little bit, the conversations began centering on what was *personally important* about the job. After all, they weren't using the money to pay the mortgage on a house with a white picket fence or to send little Johnny to college. Most of the guys did not have a traditional family, but they did have each other. Their club was their family, and they liked working with their brothers every day. They told me the demolition job allowed them to ride their motorcycles in a wide-open and beautiful part of the country, and gave them extended periods of time off during riding season. The job was more than a paycheck: it was supporting a lifestyle of independence and camaraderie. Those may be fairly unconventional values, but they are values.

After boiling it down to those simple descriptions of their values, I asked how important "independence" was as a value. Each of the men evaluated it as crucial to their very existence. I then said: "How independent are you going to be if a work injury puts you in a wheelchair? How much riding can you do if you get blinded because you did not wear your safety glasses? How much of a brother can you be to the club if you're dead?" And then I asked, "What if you simply committed to safety on the job, so you can have a richer, fuller, more reliable independence when you leave the worksite? I mean, you are living to pursue true independence, right?" Much to my surprise (and relief), the group was very receptive to those ideas. In fact, the following week they agreed to put together an employee-based "brother's keeper" safety program.

Reading about what other people value can inspire you to sharpen your own values statement. Maybe you identify with the HSE professional who has safety as a cardinal value, or the mom who works as a welder, or maybe you are more like the demolition workers. More than likely, you have your own unique way of thinking about what is important to you. The real question is, are you willing to commit to acting safely so you can continue to enjoy those things that are important to you, or are you going to continue to take risks because you are motivated by immediate gratification?

Summarizing Values

When you know what you care about, that clarity can act as a great motivator for following through on your commitment to safety. Optimizing safety requires following through on action plans, even when there are challenges. People can be tempted to skip important safety steps because it leads to immediate gratification of avoiding discomfort or feeling good right away. But being clear on what is personally important to you can help motivate you to stick to safety, even when it is difficult, because of important personal reasons.

In addition, instead of solely relying on acting safely because *the company says you have to*, you can choose to act safely because *you say you want to*! That is the big difference between compliance and commitment in the safety world. If you have not solidified why you want to be safe for personal reasons, then your safety is mostly relying on rule-following for the company's reasons. As we discussed earlier, using compliance to influence safety can be effective, but it is not optimal.

When you choose to act safely because you want to be able to enjoy all the things in your life you care about, and you have principles you want to live your life by, then your safety commitment will be stronger. When you have done the values clarification exercises, then you are more likely to follow through on action plans, see the merits of being more aware of your surroundings, and apply your skills for dealing with internal-world obstacles (which we will discuss in the next few chapters). Being able to plainly declare "What I Care About" is foundational to building safety commitment.

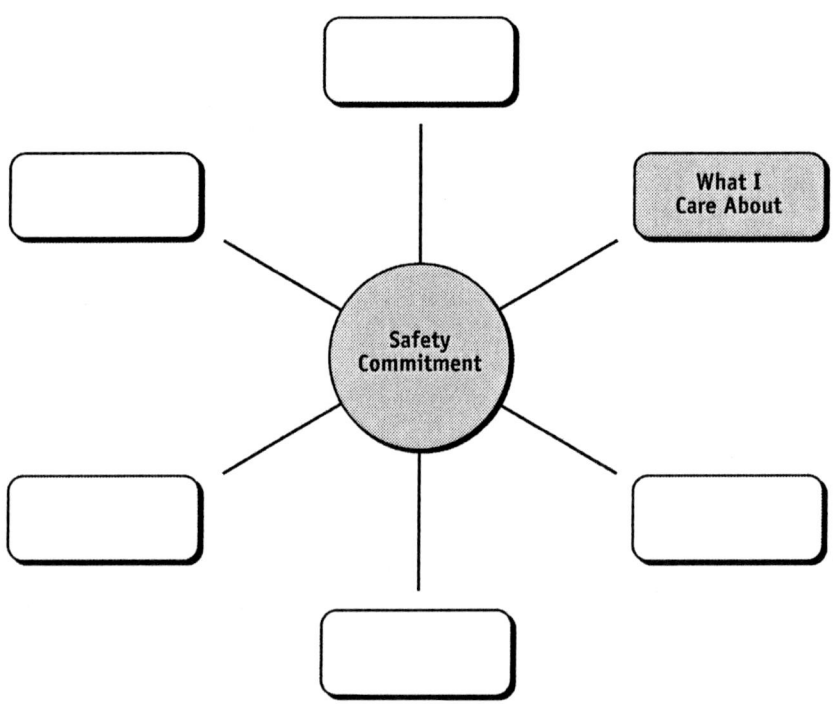

The Building Block of Your Safety Commitment Plan:
What I Care About

At the end of each chapter, you will have an opportunity to complete a section of the Safety Commitment Plan. After each part of the Safety Commitment model has been explained in a chapter, and you have experienced each of the exercises in each section, it will be easier for you to understand how each building block is integrated into your plan. In this chapter, we talked about values, and how your values contribute to the "What I Care About" section of the Safety Commitment Plan. Let's take a look at how the people we talked about in this chapter might fill out this section of their plan.

The top Health, Safety, and Environment officer:

What I Care About	Values-Based Motivation	
	--Personal Values: Describe why you work and why you aim for safety.	
	I deeply care about other people and making sure that they return home in a healthy manner.	

The woman who worked as a welder:

What I Care About	Values-Based Motivation	
	--Personal Values: Describe why you work and why you aim for safety.	
	My focus on safety at work is motivated by going home to my children. I love my family and care for my safety because of them.	

One of the demolition workers:

What I Care About	Values-Based Motivation	
	--Personal Values: Describe why you work and why you aim for safety.	
	Nothing is going to get in the way of me being with my friends and riding my motorcycle. Not even an injury. When I work, I work to have what I want, and I'll work safe so that I can keep having my freedom to get what I want.	

Now it is your turn. Use a blank Safety Commitment Plan. While writing, you don't have to use just one space like the people in the examples. Use this opportunity to examine all the different things that help you channel your energies toward acting safely.

What I Care About	Values-Based Motivation	
	--Personal Values: Describe why you work and why you aim for safety.	

Realize that you are completing one of the many parts of your final Safety Commitment Plan. Once you have completed this section of the Safety Commitment Plan, let's move forward to the next chapter about learning how to deal with some of your own personal obstacles that can sometimes get in the way of your dedication to safety.

The Second Step:

Skillfully Dealing with

Internal-World Obstacles

Chapter 5

Accepting

I'd like to start out this chapter with a confession. In the past, when I talked about "accepting" in my safety workshops, I'd get really uncomfortable. I'd feel nervous right before discussing the topic and I would have the urge to skip it because I didn't like being nervous while I was working as a trainer. I could remove the discomfort and have immediate gratification if I just didn't talk about this topic.

The reason I'd get anxious talking about this subject with audiences is because when you teach people about "accepting," you are inviting people to notice their emotions and not let their feelings be an obstacle to committed safety actions. I'd be nervous about this subject because I was afraid people would roll their eyes and think the topic wasn't cool or manly. Talking about emotions in the workplace is not a popular or comfortable topic. In spite of my nervousness, I chose to practice what I preach. I accepted my nervousness and talked about this important topic. In the presence of my anxiety I taught my audiences how to accept their own feelings –

even though I felt uncomfortable – because of my commitment to help people live safer lives. I chose to act in the direction I care about even in the presence of this obstacle of discomfort.

In this chapter, I will discuss emotions. The aim of this book is to reduce risks by strengthening each employee's personal safety commitment, which means acting safely *even in the presence of obstacles*. One of the obstacles to our safety is our habit of trying to get rid of certain emotions. Another obstacle is letting our emotions compel us to act unsafely or unproductively. Being aware of what you are feeling, accepting it, and then knowing how to keep those feelings from diverting your focus requires learning. Learning how to "accept" your emotions is a skill that helps you deal with these obstacles. Take a look at these brief scenarios, and see what happens when people aren't willing to simply accept their feelings on the jobsite.

- Kenneth doesn't like the tone of voice his supervisor uses when she tells him to make sure he uses three-points of contact when coming down the factory staircases. Kenneth feels disrespected by her, so in order to feel better, Kenneth ignores the direct order when nobody is looking. In a passive-aggressive manner, he chooses to walk down the steps without holding onto the handrail. When he rebels like this, Kenneth feels better and goes about his daily tasks.

- Fred has been working in the HVAC installation business for three decades, and at this point in his career, he gets bored very easily. He tries to escape boredom by thinking about his fantasy football team, rotisserie baseball league, or March

Madness (depending on the time of year, of course), and sometimes, he isn't fully focused on his tasks. He even daydreams when he's cutting sheet metal. Not only does his productivity suffer, but his fingertips also suffer because when he is escaping his boredom, he is most prone to cutting his hands on the jagged edges of the material.

- Christopher loved being a journeyman carpenter. He was so good at woodworking that they made him a supervisor. But being good at building things doesn't mean you are good at managing people. When he has to be the leader throughout the day, he gets nervous, frustrated, and generally uncomfortable. To manage these feelings so he can hold onto this higher-paying job, he pops painkillers he buys from a guy on night shift. The pills keep him buzzed throughout the day and allow him to avoid the anxiety he gets from giving direction to the other workers.

- Billy is fairly new to the jobsite. Despite being a rookie, he can still see his new partner, who is popular among the guys in the crew, is clearly underprepared for a work task they are both about to start. His partner is climbing scaffolding without wearing his fall-arrest harness. Billy would like to give his partner feedback about his unsafe behavior, but doesn't want to jeopardize their working relationship. Besides that, Billy doesn't want to feel embarrassed while giving feedback. He is also anxious that his new partner will give him a hard time about correcting his behavior, so Billy chooses to hold his tongue.

In each of these examples, the people are in dynamic work situations, and yet they create additional hazards by failing to deal with their emotions in a healthy way. These workers ignored rules, daydreamed, took drugs, or clammed up instead of constructively dealing with their feelings.

The Internal-World of Emotions Can Create Safety Obstacles

By the time you are an adult, you have been given thousands of messages from the world teaching you emotions should be hidden and ignored, especially at the workplace. *And that is exactly the problem!* We can get so caught up in the idea of getting rid of our feelings we don't even see *the things we do to try to get rid of emotions* can be part of the problem!

Of course, it's a good thing if you can manage your own *behavior*, especially on the job. For example, if you are angry with a coworker, you should be mature enough to acknowledge that feeling without flying off the handle, and talk about the problem reasonably. If you are sad that some of your buddies have been moved to another job site, you can still show up and work productively and safely at your own worksite. Behaving appropriately is key. Those two examples are demonstrations of acceptance and appropriate action. We can behave as we know we should even in the presence of challenging emotions. The problem is that quite often we are taught we first need to control our emotions before we can control our behavior, and that just isn't true.

We are taught early and often to avoid our emotions

We are taught early in childhood that we should control our feelings. When a child cries because he is sad that he cannot have

71

something he wants, he might hear the famous parenting line: "Stop crying, or I'll give you something to cry about!" At that very instant, the child is being taught he should control his feelings. As if a young child should not be sad when he doesn't get what he wants! The implication is that having certain feelings is inappropriate.

As the child grows older, the emotional control lessons continue. Imagine a young boy joining Little League, and during his first game, he gets his chance to bat. He starts dragging his bat up to the batter's box for the first time in his life, and the he hears his dad yell out, "Don't be nervous, buddy!" This boy has a dozen things to remember in order to be an effective hitter while he's at the plate, but the primary instruction from his father is *to not be nervous*! How exactly does a little boy stop himself from being nervous at a time like this? Is attempting to get rid of emotions really the best advice at this moment? Instead of telling the boy to control his emotions, I believe "Swing for the fences, son!" or "Keep your eye on the ball" would do very nicely as fatherly advice because they are about effective behavior. These instructions actually seem more like an action plan for effective behavior. And you can follow both of those instructions *while feeling nervous*!

Unfortunately, the instructions to deny your emotions continue all throughout life:

- Don't worry about it!
- Can't you just be happy about this?
- Don't get mad at me!!
- You're such a baby! Just cheer up and stop being so sad.
- Don't be afraid.

These common phrases are not about acting effectively, nor are they about dealing with your emotions in a healthy manner. You can see these phrases aim for emotional control, and are not talking about effective action. The point is: you can be worried, unhappy, mad, sad, or afraid, *and while you feel that way*, you can continue to behave in a committed fashion, guided by your values.

Responding with your emotions can also create risks

Emotions can also influence us to act in an ineffective manner. Sometimes we have feelings that set us up to act in an unsafe or unproductive manner. Take these statements for example: "I was too embarrassed to say anything," "I was sad, so I decided to stay home from work," or "I got so mad at my boss that I took the rest of the day off." Each phrase describes situations where emotions compelled the person to act in a risky or unconstructive manner. Let's take a look at a few situations where people are compelled by their emotions to act unsafely:

- Supervisors and operators at a manufacturing plant receive a production bonus based on meeting daily production targets. The assembly lines routinely jam and have to be cleared by the operators. Operators routinely ignore the company's lockout/tagout requirements in order to clear jams while the equipment is running. During these jams, supervisors "look the other way" rather than insist on compliance with LOTO, because that extra time will risk missing out on the bonuses. As you can see, the bonus structure is set up to reinforce the unsafe actions! And at the same time, it is obvious people have feelings of social pressure to act unsafely and they also fear angering coworkers if they decide to slow down production for safety's sake.

- Barnett is a veteran boilermaker. He doesn't get along with Paul, a newly hired safety representative. One day when they were both on a particularly dirty jobsite, Paul told all the workers to use monogoggles instead of safety glasses. Monogoggles are like lightweight scuba masks with an elastic strap that goes around the head. Paul justified this switch in PPE because the fly ash was so thick in the worksite that it was slipping past the glasses and getting in people's eyes. Paul believed monogoggles would be a safer choice than glasses. Barnett was angry at being told what to do by a newbie. He threw his monogoggles at Paul, and said loud enough for all the crew to hear: "You don't have to listen to this jerk. Monogoggles suck. Just wear what you're comfortable in." Barnett's angry outburst jeopardized more than his own safety.

- In the last seven years as a construction worker, Miguel has never witnessed anyone enacting the "stop-work authority" because of a dangerous situation. His work crew is under a tight deadline, but he believes he sees a hazardous situation brewing. It's nerve-wracking to call so much attention to himself by stopping the work process. He doesn't want to be the target of his boss's anger for missing the deadline. He is overwhelmed by the emotional discomfort related to enacting his stop-work authority. Instead of saying anything, he clams up and simply hopes for the best.

All three of those examples show how emotions contribute to workers acting unsafely. Certainly there are management issues that contributed to the risky environment, but the workers' emotions were also influential. You can easily imagine that people who are better at dealing with their anxiety and frustration might handle these challenges differently than Christopher, Barnett, and Miguel. Learning to accept uncomfortable feelings – without avoiding them or letting them take over – and then choosing to act effectively in the presence of these feelings is a building block of safety commitment.

A quick review of these examples

Hopefully you've never thrown PPE at your safety rep or taken pills to help you feel more comfortable on the job. However, if you take an honest look at your work history, I'm almost positive you'll remember times when your emotions put you in a risky situation.

If by chance you think any of my previous examples were unrealistic, then let me offer some statistics. Drug and alcohol abuse is rampant, costing companies over $500 billion per year.[8,9,10] Six percent of alcoholics and 15 percent of drug abusers show up to work drunk or high,[11] so Christopher's example of using painkillers on the job is not exceptionally rare. Barnett's aggression at work shouldn't surprise you either; according to the U.S. Department of Justice, 16 percent of all assaults occur at work,[12] and 1 percent of on-the-job deaths are homicides committed by a coworker.[13] These examples are based in reality, even if the details don't resonate with you.

On the other hand, perhaps some of these examples do resonate with you! If these examples are hitting a little too close to home for you… good! I hope it leads to a little personal exploration. Using drugs, being aggressive, or daydreaming in order to get through the day is evidence your internal-world reactions are jeopardizing your safety in the external-world. Keeping quiet when you should speak up, cutting corners when you should act safely, or being oppositional when you should be compliant with a supervisor's good advice is also more evidence you can choose better ways to deal with your internal-world feelings. Instead of letting emotions cause you to act in an unhelpful manner, you can learn to accept emotions as they are: simply internal-world feelings. Your emotions do not have to compel you to do unhelpful actions. Nor do you have to get rid of them. You can learn to behave more appropriately in their presence. Acceptance is not only a critical safety skill, but a powerful life skill, as well. Let's take a closer look at acceptance.

What is Accepting?

Accepting means *being willing to contact your internal-world experiences, without trying to get rid of them, while doing what you care about.* Let's break that definition down.

1) "being willing to contact your internal-world experiences" –

    a. "internal-world experiences" are your personal and unique emotions. When I talk about *emotions*, I am referring to feelings of anger, frustration, sadness, nervousness, and all the other types of moods that you can have on the job. As discussed in Chapter 1, there are internal and external-world experiences, and these emotional events are personal, internal-world experiences that can influence you to act unsafely.

b. "being willing to contact" means allowing yourself to be aware of your feelings, even if you don't like them. Without pushing them away or avoiding them, you simply allow yourself to feel the way you do. I did not say you could allow yourself to *follow through* on your feelings. I'm suggesting you *observe* them. This part of the definition of acceptance means acknowledging you have a feeling and not trying to get rid of it. (If you are interpreting what I'm saying as silly or you disagree with my suggestion at this point, please read on and let me explain it a bit further… we are only unpacking the definition right now. We'll talk about its application later.)

2) "without trying to get rid of them" – as you've seen throughout the examples in this chapter, people often try to get rid of "negative" feelings. Nobody likes to feel badly and we'd prefer to feel happy or other positive emotions all the time, but that is truly impossible. Trying to get rid of emotions can lead to unhealthy and unsafe actions (ex. taking drugs, ignoring a supervisor, yelling at a safety officer who is giving instructions). It is possible to observe how you feel, and while feeling that emotion, work safely and appropriately anyway.

3) "while doing what you care about" – When you have clarified your values, they can be your guidance and motivation when emotions show up as an obstacle to safe actions. Quite often when your actions are in line with your values, difficult feelings show up. For example, if you value giving feedback to another coworker so he can act more safely on the job, then you should understand that giving feedback can also be very stressful. You might be fearful of being ridiculed or ostracized for pointing out how people can take better precautions on the job. But when you make commitments to safety, you do this important task, even in the presence of your nervousness. Accepting how you feel helps you follow through on your values, and following through on your values sometimes requires acceptance.

To make this simpler, look at accepting this way: When you have worthwhile tasks to do, you don't have to *feel like* doing them; you just do them. You don't even need to wait around until you feel like working safely; you can simply work safely while accepting whatever feelings you have. You don't have to get rid of your current feelings to do the right thing; you can just do the safe actions no matter what your emotions are. That is what the word "accepting" means in this book.

Let me summarize the idea of accepting before moving on. Sometimes you won't feel like doing the safe thing, or you'll have the urge to do the unsafe thing. Accepting is about simply observing that you have these internal-world feelings, and choosing to stick to your commitments anyway. You don't let your emotions be your guide. Instead, you let your values motivate you to stick to your Safety Commitment Plan, even when you don't feel like it.

Accepting Helps You When You Are Stuck

As we've seen throughout the chapter, avoiding your own internal-world feelings can backfire. Attempting to get rid of your feelings not only doesn't work, but your attempts at doing so usually make things worse. In addition, letting your emotions override your valued actions can lead to poor outcomes, as well. The idea is to see if you can simply notice your emotions without letting them trap you.

The Chinese finger trap metaphor

Have you ever seen a Chinese finger trap? They are small tubes of weaved bamboo or wicker, and you put your left and right index fingers in both sides. Once ensnared, when you try to pull your fingers out of the tube in an effort to escape the trap, the weaving constricts even tighter. The harder you pull, the tighter the snare. The best way to deal with this trap is to fully accept that you are in it! Actually moving both fingers together – which demonstrates an acceptance of being trapped – loosens the weave and allows you more wiggle room for your actions. You get greater flexibility and an ability to release yourself from the snare. Acceptance of the situation is required.

Avoiding your feelings is a lot like trying to escape a Chinese finger trap. The harder you try to get out of an internal-world situation, the more stuck you get! For example, avoiding sadness and stress at work by playing hooky makes the work pile up even higher in your absence, which becomes something else to be sad and stressed about. Getting into an aggressive argument with a coworker because you are angry about the work environment could end up making the work environment even less pleasant.

You see, being guided by getting rid of internal-world feelings is like taking a little "hair of the dog" when you are hung-over… it might make you feel better in the short run, but what happens the next day when you are hung-over from drinking too much "hair of the dog?" When your solution to your problem becomes another problem, your trouble multiplies. Let's take a look at another metaphor for acceptance.

Monkey Trap metaphor

In a part of the world where monkeys are hunted for food, the hunters have devised an ingenious way of trapping their game. The hunters cut a small hole in a gourd, hollow it out, and tie it to a tree with a vine. Once secure, they bait the gourd with banana chips and then hide in the trees nearby. When the unsuspecting monkey sniffs the scent of bananas while foraging for food, he eventually finds the source of the scent and slips his hand into the gourd to grab the banana chips. When he does so, the trappers jump from the trees and run toward the monkey. Panicked, the monkey tries to pull his hand from the gourd, but because he has a handful of banana chips in his clenched fist, he cannot pull his hand out. He tries to run to avoid the hunters, but he just winds himself around the tree.

If he would "just let it go" and accept the loss of these banana chips, he could then behave more flexibly, and slide his hand back out of the hole. But his unwillingness to lose that banana chip creates a problematic situation, and ultimately leads to him being dinner for the hunters.

This story simply serves as a metaphor about how you can best deal with troubling emotions. People grab on tightly to their own agenda and want things to go their own way. Sometimes it is better to stop trying to control your internal-world feelings and instead focus on your commitment toward finding safer and more valuable banana chips elsewhere. Rigidly pursuing a single-minded agenda of insisting that things go your way can end up being unproductive and unsafe. This is especially true for anger. At times, we get so angry about things not going our way that we clench our fist, and even our minds, around an idea, and struggle to make sure we hold onto that ideal. That struggle can be unproductive and backfire. To put it simply: sometimes you should "just let it go!"

This banana chip story had an impact on the safety culture at a mid-sized construction company that participated in one of my safety training workshops. The CEO of the company explained that when one of his crews was working overseas, a late shipment of building materials held up progress on their job. A crew leader and his supervisor were viciously arguing whether they should knock off early and go back to their hotel because of this problem. The crew leader was infuriated, and said, "I'm sick of being disrespected by this shipping company! They never seem to deliver on time and it screws up our timeframe! Me and the guys are just going back to the hotel!"

The supervisor answered back, "Just because we don't have that shipment doesn't mean we don't have any work to complete. Do not leave! That will mess up the deadline, and then all the guys will be rushing to finish the job at the end of the week. That'll create unsafe conditions. Just start doing the other phases of our work plan!"

The crew leader refused because he and his workers felt angry and disrespected. They were also irritated that job completion was going to be delayed and it might take longer for them to go home.

Finally, one of the workers spoke up and said to the crew leader: "Hey man, just drop the banana chip! We'll just let that stuff go, and get more productive on the other jobs we *can* do! Just because you're angry doesn't mean you can't be productive! Just let it go!"

The crew leader became more levelheaded and the gang refocused their attention on even-keeled productivity. In fact, they accomplished a great deal before the shipment finally showed up. The workers were in contact with their feelings of anger, but still chose to act in line with their values and commit to working productively. Because of their diligence, nobody had to rush around at the end of the week to make deadline, and they all went on to enjoy safer "bananas."

A Mint Condition Acceptance Exercise

I'd like to invite you to try an exercise that will help you understand a little more about accepting, and maybe even strengthen your acceptance skills. When you have three minutes, see if you can get a hold of a small piece of hard candy. I'm not kidding… the exercise is really about eating candy! (A peanut, mint, or raisin will work just fine, but I'm using hard candy in the example).

Read through the rest of these instructions before actually trying the exercise. Try to remember the instructions as best as you can because you can't read the exercise and do the exercise at the same time. You can also listen to an audio recording of this exercise to help guide you through what you should be doing. (See Appendix B for resources for obtaining this exercise on audio recording.)

Once you are in possession of candy, get ready to eat it. But I'd like you to do something differently than you usually do with this candy. As soon as you pop it into your mouth, let it rest right in the middle of your tongue. Center it in the middle of your mouth. Take a nice, full clean breath, and as you exhale, close your eyes (or affix your gaze on something around you). Observe the piece of candy sitting on your tongue. See if you can detect what it feels like there. Observe its shape. Observe its texture. Can you detect how you are reacting to the candy? Can you feel your mouth water?

Now, notice some other feelings. Can you feel your urge to swirl that candy around? Do you feel like chewing on the candy? See if you can be aware of any feelings and urges, and commit to keeping that candy right on the center of your tongue because you care about expanding your acceptance skills. Take another full, clean breath and refocus on what it feels like to have that candy in the center of your tongue. Observe if it feels odd to just let candy sit there without sucking on it or rolling it around in your mouth. Commit to keeping that candy there even if you feel like moving it. Accept those urges and those feelings. Come into contact with how you are reacting to this exercise, and don't try to get rid of those feelings. Don't move the candy just so you feel more comfortable. Don't distract yourself from the experience either. Simply observe how you feel, fully and without trying to get rid of the discomfort. Observe the frustration that might have come along with this new experience. See if you feel like giving into your old habits and just notice that urge to fall back on old behaviors. Maybe you want to give in to your old routine of putting the candy in your cheek. Instead of doing old habits, simply accept that those internal-world feelings are with you, and maintain your commitment to the exercise.

Debriefing the exercise

Welcome back! You can chew the candy now if you want to. The main point of the exercise was to help you strengthen your ability to observe your own feelings and urges without doing anything to get rid of them or respond to them. This simple exercise demonstrates the possibilities. People can observe their internal-world experiences without giving into them. You can simply notice the urge to chew on the candy while behaving differently. Your internal-world does not have to cause your actions.

Do I expect you to do this exercise one time and never behave impulsively due to your emotions again? No! If we ask Kenneth, Christopher, or Barnett to do a three-minute exercise with a piece of candy, are they going to suddenly become rational and fully safety-minded? Of course not! That would be like suggesting you do a few ten-pound barbell curls to build up your biceps to make you stronger while working on your construction job. In order to get stronger with accepting, you need repetition and practice. The question you need to ask yourself is, "Am I willing to engage in the safe actions, even when it's emotionally difficult or when I don't feel like it?" That is how you get you "reps" in, and that is how you build up your "acceptance muscles."

Please Do Not Distort the Meaning of Acceptance

Accepting is about acknowledging your feelings and emotions while making sure you stick with your values-based commitments. Please do not misinterpret this concept. It is a total misunderstanding to think if you are around a welding arc, and you flash burn your eyes, that you should be willing to notice this feeling and continue working in the presence of this pain. That wouldn't be accepting... that'd be stupid! If you get welder's flash, you should continue with your commitment to behave in a way that helps you maintain your health... by getting medical attention!

By the same token, suppose your boss tells you to work on some electrical lines that have not been locked and tagged out. Of course, you would refuse by saying "Boss, I don't really feel comfortable working without locking out." If he turns around and says, "I don't care if you feel like it or not! Accept the fact that this is your job and just do it!" then he has totally distorted the idea of accepting. That kind of attitude has caused many injuries and cost many lives and should not be tolerated.

Please understand, I am not suggesting you accept things you *can* change. If you can alter things in your external-world, by all means do so! If there are external-world safety hazards, do not accept them! Fix them! And do everything you can to ensure that the company addresses the external-world issues. Keep in mind, acceptance is not about allowing problems to simply exist in the external-world. Acceptance is about learning how to deal with obstacles in your internal-world.

The Serenity Prayer is a widely popular piece of wisdom that helps people deal with external and internal-world problems. It says, "God, grant me the serenity to accept the things I cannot change, courage to change the things I can, and the wisdom to know the difference." This phrase helps people all around the world who are struggling with lots of different challenges, and it can be applied to your own safety commitment! Aim to develop skills to change external-world problems, while having the abilities to accept the internal-world feelings that you can't change.

The Building Block of Your Safety Commitment Plan: *Accepting*

In this chapter, we talked about allowing yourself to acknowledge your emotions without trying to control them, and being wiling to simply have those emotions while performing safely and productively. Demonstrating this attitude on the job will contribute powerfully to handling internal-world obstacles while executing your Safety Commitment Plan. "Accepting my emotions" is part of the commitment phrase so you remember to enact the skill of accepting how you feel before moving forward with your work tasks.

Accepting	Allow yourself to acknowledge any emotions you are having without trying to control the emotions. Be willing to simply have those emotions while moving forward with safe and productive actions.

Chapter 6

Noticing

In this chapter, we explore how distracting thoughts can cause safety problems, and how you can use the skill of *noticing* to help you deal with these dangers. In the last chapter, I explained how emotional reactions could become obstacles to keeping a commitment to safety. And if emotions are thorny problems to deal with, then thoughts can be a whole briar patch! You see, your emotions change a few times each day, but your thoughts change moment to moment. If you respond to your own thinking process in the wrong way, you can potentially slip-up on your commitments at almost anytime. Sometimes you can get "caught in a thought" and be misled by your own mind to do something risky. Believe it or not, your own mind can occasionally act as an obstacle to safety.

I'm not trying to criticize the human mind. I am personally glad I have the ability to use my mind to plan things and solve problems. I imagine you feel the same way, too. Being able to think your way out of a difficult situation is one of the greatest human abilities. But here's the problem: even sensible, educated, and mentally healthy people still have to deal with the fact their mind occasionally gives out erroneous information and bad advice. Our mind quickly and strongly influences our actions, and can also quickly pull us away from our safety commitments. Even further, we don't always have our "mind on task," and the skill of noticing can assist with these problems.

The Power of Thinking

Our thoughts strongly affect how we behave. This is probably why one of the most popular slogans you see posted all over jobsites is "Think Safety." Because our thoughts are usually helpful to us, we trust them and usually follow through on what our mind says. For example, if you awake in your home to an intense odor of natural gas, your mind may say: "There's a gas leak in the house. This is unsafe. I better get out now!" You don't need to have direct experience with natural gas explosions to behave effectively. Instead, the thinking process immediately rouses you to hightail it out of the building before it ignites. Your mind helps you act safely without always having to learn from past mistakes and painful experiences. Thankfully, your mind can make quick and strong responses to problems in your external-world!

But in some situations, *thinking* can be the problem. This is especially true when unhelpful thoughts quickly and strongly influence us to act in a dangerous manner. Here are a few examples of thoughts that trip up safety commitment:

- "I feel nervous about giving feedback, and I hate this feeling. I'm better off just staying silent during the toolbox meeting so the feeling goes away."
- "This safety inspection project is annoying. I'd rather go back to my office and check my emails than work on this inspection."
- "I'm sick of this supervisor's attitude and I ain't listening to anything she's got to say!"

Those are three thoughts that could show up during the workday, and since our mind has so much influence over our behavior, these thoughts might lead to potentially hazardous actions.

Obviously, those are just simple examples, so I invite you to check in with yourself and see if you can recall any unhelpful thoughts you have had lately. Have you had thoughts leading to you toward performing risky behavior? What thoughts prevent you from doing your best work? Insert your problematic thoughts here.

_____

_____

_____

If you couldn't think of anything to write, try to pay attention next time you are working, especially when things aren't going well, and see what you are saying to yourself at those times.

Our mind helps us respond to our world in a fast and forceful manner, and for the most part those are useful qualities of the mind. But there is a dark side: the *quickness* and *strength* of our mind can lead us to act impulsively, rigidly, and unsafely. The quickness and strength of the mind are at the core of the problem of getting "caught in a thought." It's like our mind sets us up with a one-two punch when these unsafe thoughts pop up. The lead punch of this combination is so speedy, we don't even realize what hit us. Our mind works really quickly, and we can barely do anything to stop it.

Quickness of the mind

Our mind's capacity to rapidly process information is a valuable asset. This ability helps us in so many ways. Suppose you need to order a dozen pallets of material to meet a demanding deadline. Your vendor says, "Each pallet costs $1000, and all twelve will cost you $14,000." Your quick thinking abilities let you figure out your vendor is overcharging you $2000. This ability comes in handy before you sign a purchase agreement. At another time, when your boss says, "What was the name of that customer who helped us make our quota last year," and you can quickly answer "That was the Frisone Corporation, sir," you've benefited from being able to instantly relate two important things in your world: your boss's question and the answer he was looking for. We benefit greatly from our quick ability to think about our world. But sometimes that quickness can cause trouble. I invite you to please stick with me during this next fun exercise about the speediness of your mind.

You see, your mind works so rapidly that I know exactly what you're going to think after you read the next sentence. Even though you and I have never met, and you are a perfect stranger to me, I can make your mind react a certain way… ready? "Mary had a little _____."

See, I know your mind just said "lamb. " You can deny it if you want to, but if you learned to speak English as your native tongue, then I'm betting that your mind immediately said "lamb" after "Mary had a little _____." Your internal-world has been conditioned to think about certain things when the external-world gives certain signals. It's extremely difficult to stop yourself from thinking about certain things. Read this next line to yourself and try not to think "lamb."

Mary had a little _____.

You thought lamb. I'm positive you did. Perhaps, you might have tried to suppress thinking "lamb" by not looking at the phrase, but then you are no longer working on the task! You might have tried to replace the word "lamb" with something else, like "skirt" (and that just goes to show what other kinds of thoughts are automatic), but in the end, once the "lamb" connection has been ingrained, it is not easily replaced.

Even if you insist you were somehow entirely successful at not thinking "lamb" during that last time I presented those words, it's unlikely you'd be able to keep that mind-control mechanism in place forever. If you weren't expecting it, someone could come up to you tomorrow and say, "Mary had a little _____," and you would just automatically think, "lamb." This is not some kind of mental weakness on your part.

When certain things happen in our external-world, it is often helpful for the internal-world to respond rapidly and automatically. But it isn't always helpful because hazardous thinking can instantly pop up in the same way. It's possible a looming deadline makes you think about cutting a few corners. Sometimes you get distracted by thoughts about your family, finances, or your favorite sports team. Other times your internal-world reminds you of something you learned in the past but is not helpful in your current work task. In all those situations, your thoughts are causing trouble. As I said earlier, getting "caught in a thought" is like a one-two punch. The first punch is how automatically and quickly your thoughts come at you. Punch number two is a powerful right-cross: your thoughts can have a significant impact on you!

Strength of thoughts and words

The words you say to yourself when you're thinking should have some influence on your action, otherwise, what is the point in thinking at all? The problem comes when we rigidly obey every thought that comes to mind. It's problematic when we act like robots that follow the mind's programming. Check out this next demonstration about how the words you say to yourself influence your reactions.

Can you imagine holding a ripe lemon in you hand? Right now, can you see yourself holding one of those giant yellow lemons from the produce market? You can smell the lemony scent and can tell it's really juicy inside. The skin of the lemon is starting to stretch because of how much juice is on the inside. You can almost taste the sourness as you place it on your cutting board. Your sharp knife slices through the lemon and juice sprays out and pours all over the cutting board. The two halves of the lemon start rocking back and forth in the puddle of sour juice leaking from the fruit. You grab one of the dripping halves and bring it close to your lips while smelling the fresh lemon scent. At first, you sip out some of the juice and the sides of your tongue tingle. You then squeeze the lemon juice right into your mouth and feel the tartness on your tongue while swallowing the sour juice. You squeeze the lemon again and drink down the juice. Then you go in for a second bite of the lemon and chew on the sour fibers while licking the inside of the lemon peel.

Are you salivating right now? Maybe even just a little bit? If you are like most people your mouth starts to water even after *just reading* about lemons. While you were reading that paragraph, did you actually drink any lemon juice in order to stimulate your salivary glands? Of course not! The only things present here are the mere words on a page. But you've learned that these words have meaning, and therefore have enough *power* to get your body to start reacting as if lemon juice was actually put in your mouth.

Your mind is strong enough to present words and images that influence your actions. The mind's ability to act quickly and strongly is helpful sometimes, but dangerous at other times. When the strength and quickness of your mind lead you into trouble, you are "caught in a thought."

Getting "caught in a thought"

When your internal-world makes you interact with the external-world in a way that isn't good for your desired goals, you've gotten "caught in a thought." It occasionally happens to everyone and in different situations… not just in the workplace. Our thoughts jeopardize our safety by causing us to act in lots of unconstructive ways:

- Impulsively - "I can just skip this safety step to save time."
- Avoidantly - "I'm not wearing those stupid-looking monogoggles"
- Irresponsibly - "Safety is the safety representative's job, not mine!" or "No one told me I needed to report that problem!" or "I'll just send one more text message before I reach the highway exit ramp."
- Arrogantly - "I've been doing this job for 10 years. I haven't been hurt yet, so I'm not changing what I'm doing!"
- Ignorantly – "I don't need to learn anything new about safety."
- Complacently – "It won't happen to me." or "My six year-old won't get hurt riding his bike around the block just once. No need to make him wear a helmet."

These are just a handful of distorted thoughts that all come from the internal-world, can quickly exert power over our actions, and may become obstacles to optimum safety.

Noticing Your Thoughts

If you combine all the things we've said so far in this chapter, you see why I said thoughts are like a briar patch. We can get "caught in a thought" lots of times throughout the day. Our mind often gives us thorny messages that stick with us and lead us away from our commitments. So you might be wondering what can you do with these problematic thoughts? My short answer to that question is: *Nothing!* (The long answer is what I'll be talking about for the rest of this chapter.)

When I say you should do "nothing" about your problematic thoughts, I'm humorously implying you should *not follow through* on the thought. Just because you have an unconstructive thought does not mean you need to do what it says! You are not a slave to your mind. Just because your mind feeds you some instructions does not mean your actions have to obey. When you have a well-developed Safety Commitment Plan that is motivated by following through on your values, you don't have to let a random thought throw you off of your commitment. You can learn to distance yourself from unhelpful thoughts, and that distance might give you enough time between *thought* and *action* to make more values-based decisions instead. If you can learn to simply *notice* that you are having thoughts, rather than be dragged along by them, you will prevent yourself from impulsive, avoidant, and irresponsible behaviors.

Noticing your thoughts is a powerful skill, and it takes a little bit of time to develop. With practice, you can acquire an ability to distance yourself from your thoughts so you can get a better perspective on what you are telling yourself. I'm inviting you to change the way you interact with your internal-world. It takes some getting used to, but people who are good at observing their internal-world become more in control of their behavior, are more skilled at keeping their "mind on task," and are better at sticking to their commitments.

Exercising the separation between thought and action

Let's try a simple exercise: Keep reading this page, and also wiggle your toes for the next 25 seconds without stopping. Commit to that behavior. Don't stop no matter what you read next. Maintain your toe-wiggling commitment. Begin now.

As you wiggle your toes, say to yourself, "I can't wiggle my toes. No matter what I do, I cannot even move my toes." Repeat it three times.

Did that simple statement stop you from moving your toes? Probably not. I imagine you kept on wiggling your toes. As I said throughout the beginning of this chapter, your mind can have a powerful impact on your actions, but what I'm telling you now is: *it doesn't have to!* You can create a distance between your thoughts and your behavior. (On the off chance you did stop wiggling your toes when you were telling yourself "I can't wiggle my toes," just try the experiment again. I am positive that now you know the aim of the exercise, you'll be able to do it).

Try this type of experiment again with different actions. Tell yourself, "I can't raise my right hand," and put your hand in the air. Tell yourself, "It is absolutely critical that I snap my fingers three times right now," and notice that your mind is saying that phrase while your fingers remain perfectly still. Hear yourself privately say, "It's too difficult for me to tilt my head back and forth," and observe your mind saying that thought while using your neck muscles to rock your head back and forth. Thoughts do not have to cause actions if those actions are unhelpful or dangerous. The skill of noticing is similar to the skill of accepting. Just because you have a certain thought or emotion does not mean these internal-world events have to sway your behavior to become unsafe. Neither thoughts nor emotions have to interfere with values-based actions when you are committed to doing them. The combination of noticing and accepting can be a powerful tool for dealing with internal-world obstacles to safety.

A sour example about noticing

Your thoughts can be powerful, but they don't have to be. To remind you how powerful words can be, think back to that description of a lemon, and how you started salivating to those mere words. Now read this next paragraph to the very end. (If you skip over it or tell yourself that you "get it" without actually reading the paragraph word-for-word, you will not actually "get it." The exercise is worth your 30-second investment.)

Lemon lemon

You just read "lemon" 80 times. Earlier in this chapter, you were probably salivating while you read about drinking fresh lemon juice. Well, after reading the word "lemon" 80 times, you figure you should be salivating so much that you start to drool. But I'll bet anything you were not salivating by the end of that paragraph. If you truly read every one of those 80 words, I imagine you stopped thinking about a round, yellow citrus fruit about halfway through. Heck, I bet you weren't even picturing a lemon in your mind when you were about halfway through that exercise. Even though that was an onslaught of "lemons," after a little while the words you were reading had very little meaning to you. The words likely just became sounds in your mind: *lemonlemonlemonlemon*. Or maybe you just saw a blur of black squiggles in your line of vision. The words you were saying to yourself lost their meaning.

Ponder that idea for a second. It is important. The words you say to yourself *can* lose their impact. It bears repeating: the words you think in your internal-world can lose their influence on you. When you think something, those thoughts do not have to cause action. This is true for a great majority of thoughts: you can simply notice your thoughts are happening without necessarily letting them influence your behavior.

Our minds get conditioned to automatically think certain things (Mary had a little _____). It is also difficult to change how quickly your mind reacts to the world (2+2= ____). *But you can develop the skill of reducing the power of your unhelpful thoughts on your actions.* Simply start to observe that your mind is like a word machine and it is overproducing words all day long. You might not always be able to control what thoughts your mind throws at you, but you can decide how you will respond to those thoughts. The point is, just because your internal-world generates an unsafe thought, doesn't mean you must follow through on it. The impact of your words is reduced when you learn to distance yourself from the thought by simply noticing they are happening. Let's take a closer look at developing skills for noticing thoughts.

Practice Noticing

If you'd like to get better at distancing yourself from unhelpful thoughts, find some time everyday to simply notice your thoughts while doing something you typically do. If you drink coffee everyday, take a minute to *just do that*. Simply commit to drinking your coffee. If your mind starts telling you to do something else, notice your mind is talking to you. Distance yourself from those thoughts. Simply commit to drinking your coffee for one minute a day without getting caught up in your mind's distractions. If your internal-world tells you to check your emails, read the paper, or daydream, see if you can observe that your mind is demanding you do something besides your commitment, and then simply refocus your mind and do the one thing you set out to do: drink your coffee.

If you aren't a coffee drinker, then you can practice staying committed to almost any simple behavior while noticing your thoughts. For instance, take a short walk while simply noticing your mind chatter. See if you can calmly hear that inner voice of yours. Observe it. And as you walk this short path for this minute in your day, see if you can just *hear* your inner voice, rather than *listen to* your inner voice. This exercise can help you start to notice how quickly the mind generates thoughts, and helps deal with how strongly they can influence you. When you start to notice thoughts, rather than obey your thoughts, you can develop a skill for guiding your actions by your values and not what your mind automatically comes up with.

Of course, when you're practicing noticing by doing this walking exercise, please watch where you're going! Even though you are learning to distance yourself from the quick and powerful automatic thoughts, you should plan to walk safely, and your plan should be motivated by your values.

A Mint Condition Noticing Exercise

It's time for more candy! I'd like to invite you to try this exercise again by popping another mint into your mouth (a peanut, raisin, or any kind of hard candy will still work just fine). Center that mint in the middle of your tongue. As you keep it there, watch your mind start jabbering away. Once you commit to letting that candy sit in the middle of your mouth, notice how your mind gets active about the mint. Notice if it starts to rebel against the instructions from this exercise. Just observe the thoughts, and make sure you don't follow through on the thoughts for three minutes. Your mind may tell you lots of things during this time period, but don't stray from your commitment to keep the mint directly in the center of your tongue. Stick with it, even if you think:

- "I want to chew this."
- "Man, this is a boring exercise."
- "Why do I have to leave food in my mouth?"

- "This is a good tasting mint."
- "I wonder if this candy is artificially flavored, or if it comes from real spearmint leaves?"
- "I've had enough of this stupid exercise."

How did that exercise go for you? Were you able to simply notice your mind was telling you things while you maintained your commitment? Did your internal-world start to protest against the instructions? Hopefully, you were able to notice those thoughts while maintaining your commitment to keeping the candy where it was supposed to be. Maybe your mind just wandered and you started thinking about something else, and you simply forgot you were in the middle of an exercise. That can happen too, and we'll address how easily our mind wanders in the next chapter. The purpose of this exercise is to have firsthand experience on what it's like to simply observe your thoughts without obeying your thoughts.

Noticing what your mind says - without following through on what it says - can be difficult at first, but with practice you can get better at this skill. If you would like to improve your skillfulness with noticing, see Appendix B for resources to obtain this exercise on audio recording, and use it to keep practicing! When you get better at distancing yourself from unhelpful thoughts, you will be better at building safety commitment.

Thoughts versus Values

Creating distance between your unhelpful thoughts and your actions will certainly reduce your risk for injury. But you still might be wondering, "If I don't listen to my mind, what's supposed to guide my actions?"

Safe actions should be guided by guidelines from your company, OSHA regulations, and professional job hazard analyses. There are many technical and procedural approaches beyond the scope of this book that should govern your behavior. The purpose of this book is to help your commitment to those processes and procedures. Proper professional training, observation, and feedback should guide your actions, and a well-developed Safety Commitment Plans can integrate your training and experience, company policies, your personal values, and the other skills learned in this book. We will talk more about how specific safety steps should be integrated into your Safety Commitment Plans in Chapter 9 called "Doing."

The best Safety Commitment Plans are based on proven methods for keeping you safe. The guidelines that have been proven effective to protect you on the job should guide your actions. When your internal-world tells you to wear a hard hat, by all means follow through on that thought because OSHA and your previous training have influenced you to think those thoughts! "I should put a hard hat on," is not a random thought that just pops in your mind; it is based in keeping you safe and is linked to your commitment plan. Thinking about putting on your hard hat is parallel with your commitments. It's also in-line with what you care about.

If you think, "I should put a hard hat on," while walking into an industrial site, *do not just notice that thought!* Obviously, that thought is in line with your commitment plan and values, so it should influence your behavior. Deviations from commitment plans are often the cause of injury, and often the reason for deviating from a commitment plan is because the worker got "caught in a thought."

Building safety commitment comes from working on two opposite sides of your internal-world: your values and your thoughts. On one hand, safety commitment is strengthened by *clarifying* what is personally important to you to motivate you toward greater safety actions (see Chapter 4). On the other hand, your commitments will be stronger when you become good at *noticing* thoughts that are inconsistent with your values so you don't engage in unsafe actions.

It is impossible for a book to tell you exactly which thoughts to notice and which to follow. Your mind will throw lots of different thoughts at you all day, and I can't predict what they will be. But when the thoughts are out of line with what you've been trained to do, it's a good idea to simply notice those thoughts. If the thoughts compete with what you truly value, don't follow through with it because it won't serve what you truly care about. If the thoughts are motivated by your emotions or impulsivity, see if you can accept the feelings you're having, notice the thoughts, and behave safely, even in the presence of these obstacles.

The Building Block of Your Safety Commitment Plan:
Noticing

In this chapter, we talked about how your mind can sometimes act like a word machine on overdrive. It just keeps churning out a stream of thoughts... some are helpful, some are not. You can strengthen a skill called "noticing" that allows you to continue acting in the safe manner, even if your mind is blabbering and pulling your attention away from what you are doing. As you review your Safety Commitment Plan checklist before starting your task or beginning your workday, you will remind yourself of this critical skill: simply noticing your thoughts as they arise, and letting them go if they are not helpful.

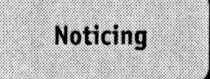	Prepare to simply notice thoughts that arise during the work task. Let them go if they are not helpful. Treat distracting thoughts as disconnected from action while choosing to act in a meaningful and safe manner.

Chapter 7

Here Now

In the last two chapters, we looked at how thoughts and emotions can sometimes obstruct optimal safety and take our focus off our commitments. Our internal-world can distract us and put us in jeopardy of getting hurt by external-world hazards. Commitment to safety requires on-going vigilance. To say it another way, optimum safety comes from concentrating on what is happening in the present moment, and constantly focusing on what is going on "here and now" in your workplace.

Yet, research shows that people's minds are not focused on what they are doing for about 47 percent of the day![14] Almost half of the day, we're all thinking about something other than what we're doing. Consider how that fact contributes to the risk of injury in the workplace, and imagine how much better off we all would be if we could reduce the level of distraction for each worker in dynamic worksites. Think about how many commitments would be kept if our minds didn't wander off so easily and frequently.

If you are skeptical about the fact that almost half the day people are thinking about something other than what they are doing, I invite you to simply observe people's behavior. It won't take you very long to see how quickly people become distracted. It seems people's minds are often not focused on what is happening "here and now." Instead, people are often thinking about other things than the world right in front of them. Take driving as an example. Many drivers seem to be doing almost anything but paying attention to the road. When I look at people behind the wheel of a car, I see a surprising number of drivers texting, talking on the phone, eating, applying makeup, or some other action besides safely maneuvering the vehicle. And in addition to the obvious distractions, every driver is also susceptible to daydreaming or mind-wandering instead of concentrating on the road.

Have you ever had your mind wander while you were driving down the highway? Have you ever been so lost in thought you drove right past your exit ramp? Whenever I ask people in a safety workshop to raise their hand if they've ever zoomed past their highway exit because they weren't paying attention, I almost always see 100% of the people raise their hand (and the people who don't raise their hand were probably not paying attention to the question!)

Do you realize that in this kind of situation, your mind is an obstacle to optimum performance and safety? We sometimes get so distracted by our thoughts we miss our cue to exit *even when the exit signs and lane markers are in place and well-known to us*. The external-world distractions cannot be blamed for that kind of mistake; rather, the internal-world is the primary cause of this safety problem. Our minds get led away from paying attention to what is happening here and now and put us at-risk for neglecting important signals for action.

Driving by your exit is just one example of how your internal-world can interfere with your situational awareness, and lapses in awareness can put you in worse jeopardy than just accidentally missing your exit ramp. Working in a dynamic environment requires vigilance because the workplace can present multiple threats to your safety. The problem is that distractions from your internal-world occur almost all day. Thoughts, emotions, memories, daydreams, making plans for the future, and other private experiences happen throughout the day. These events don't always lead to injury, thankfully, but they do diminish optimum safety, especially if you really get caught up in them. This lack of paying attention to what is happening right here and now is a potential hazard.

Injuries can occur when you aren't in contact with the present moment. People do not like to admit these kinds of things in official incident reviews, but after injuries I have heard people say:

- I lost focus for just a second.
- I wasn't really paying attention.
- This part of the job is so boring that I was just kind of going through the motions.
- Well, I had a lot on my mind and I just wasn't ready for this kind of thing to happen.

These phrases describe moments when the worker was not really in the here and now, and because of that momentary lapse of attention, got hurt. The purpose of this book is to reduce the number of people who get hurt by building safety commitment. The purpose of this chapter is to improve your safety commitment by increasing the amount of time you spend focused on the here and now.

The Only Time You Can Act Safely is Right Now

Most of use see the merit of having our mind on task, and in the here and now. Since research shows that almost half of your waking hours are spent thinking about something other than what you are doing, the potential risk that comes from your internal distractions should be obvious. While I understand it's natural - and even beneficial - for you to occasionally think about the future or the past, it's best to focus on the present moment when you are engaged in a work task.

The only time you can act safely is right now. "Now" is the only time you are engaged in action. "Now" is the only time you can behave safely. You can't be safe tomorrow and you can't be safe yesterday. (If you think you can be safe tomorrow, prove it! In 24 hours, act safely and ask yourself if you are doing it "now" or doing it "tomorrow!") You can't wear your hard hat in five minutes, and you can't put it on five minutes ago. "Now" is the only time you can wear your personal protective equipment. (I have met too many people who wish they could have put their PPE on just five minutes before their injury). The only time you are able to act is right now, but your internal-world isn't always focused on what is going on right now. This chapter aims to help you learn to focus your mind on the here and now.

Situational awareness

In recent years, safety professionals have seen how important it is to reduce distractions at work, and have also shown interest in "situational awareness" as an important characteristic of behavior on the job. Situational awareness has been defined a number of different ways,[15,16] but for the purposes of our discussion, situational awareness is the perception of external-world and internal-world elements that can influence your behavior. Even more simply, situational awareness is knowing what is happening in your environment so you know how to act appropriately.

I believe we sometimes take situational awareness for granted. Situational awareness is a skill that needs to be sharpened. Many safety workshops only focus on teaching how to detect unsafe situations on the job and what to do when a hazard is detected. However, we often overlook teaching workers the *skill of staying alert* so they can execute the other two skills. When your mind wanders or gets caught up in thoughts and emotions, your ability to detect dangerous situations is weakened. To make an analogy, you can train people the right way to swing a hammer, but they need to exercise their arm muscles to be able to do it strongly and correctly. By the same token, you can train people to detect safety hazards, but they need to exercise their mind to strongly and correctly maintain their focus on detecting these hazards, as well. The next section offers some of these exercises.

Safety commitment requires a strong ability to focus, refocus, and refocus again on what is going on in your environment in the present moment. There are evidence-based methods that you can use to sharpen your focus on the being in the here and now.

Improving Situational Awareness

Practicing the skill of being in the here and now has been shown to have remarkable effects on health and well-being. The more often you are focused on what you are doing, the more likely your actions will lead to optimal performance. We will discuss a few different exercises to help you become more present-focused. These exercises will sharpen your focus, reduce stress, improve memory, and stick to your commitments for a longer period of time.

The main purpose of situational awareness exercises is to extend how long you can pay attention to what you are doing. If you increase the duration of your attention span, you are more prone to detect obstacles that might get in the way of your commitments. Improving your contact with the present moment helps you detect external-world hazards. Increasing your mindfulness also helps you be present with your internal-world obstacles while continuing to behave in the direction of your values.

Take a breath

Learning to be in the here and now is fairly simple, but it is not easy. One of the most popular ways you can learn to stay in the here and now is to simply learn to pay attention to your breathing. This training method has been helping people become more present-focused for centuries. In the 21st century, millions of people continue to practice paying attention to their breathing as a way of increasing performance and well-being.

Here is a simple invitation. For your next ten breaths, just notice the experience of each inhale and exhale. As you inhale, notice the sensation and simply observe the feeling of breathing in. As you exhale, observe what that feels like. If a thought arises, or you start judging whether you are doing it the right way, just let those thoughts drift away as you refocus on the sensation of your breath. If you get an urge to move other parts of your body, or you feel a twinge in your neck, or an itch on your knee… simply notice them.

Remembering what you learned in the last chapter: you can't stop your thoughts but you also don't have to follow through on them. Simply be present with them and still stay committed to doing one thing: paying attention to your breath. If your mind wanders, then when you catch yourself doing that, guide your attention back to what you are doing: paying attention to your breathing. Now that you have read through the instructions, please give this a try.

Whenever you do this exercise, thoughts, feelings, and urges inevitably appear. Your mind might start making judgments that you aren't doing it correctly. Having these internal-world distractions does not mean you are doing the exercise wrong. Your mind is constantly active while you are awake, and does not turn off easily.

Because you're constantly breathing, focusing on your breath is a simple introductory exercise for strengthening your situational awareness. Your body is either inhaling or exhaling at all times. The exercise just asks you to notice it. Learning to be more aware of your breathing creates a tether line between your internal experiences and the current moment. You are always breathing *right now*! If you practice this exercise daily, you will not only become more in-tune with your breathing, you will be more fluent at contacting what is going on in the present moment.

Remember, the purpose of this exercise is to strengthen your ability to focus, refocus, and refocus again. The more you do the exercise, the stronger your ability to stay attentive to what you are doing. Ideally, this exercise (and the ones that follow in the next few pages) will increase the duration of your attention span. When you become distracted, get "caught in a thought," or wrapped up in stressful emotions, taking a breath can help you get more centered and situationally aware of your surroundings. From this stance of greater awareness, you can be more accepting of your emotions, notice your thoughts rather than get hooked by them, and refocus on your value-directed commitment to act safely in the workplace.

A Mint Condition Here Now Exercise

Grab your mints; it's time for another exercise. You might be thinking, "More candy?" And my reply is: Yes, let's do the mint exercise one more time. The mint exercise is an effective way to learn how to be in the "here and now." (I imagine this is the first book that aims to improve people's personal health by encouraging eating candy. If you want to use something healthier like a raisin or a peanut that will also work.) This is the last time we'll do the mint exercise in this book, so see if you can truly commit to following through on the instructions.

Put the mint in the center of your mouth and simply commit to keeping it in the middle of your tongue without moving it. And now become aware of the situation. Be fully present with what you are experiencing while observing that your mind is throwing lots of thoughts at you. Notice the shape of the candy on your tongue. Is it round or square? Is it rough or smooth? Notice the texture and the flavor. If your mind starts to wander or judge the exercise, simply refocus on the present situation: what it feels like to have that mint in your mouth.

Observe the way your body is reacting to the mint right here and right now. Maybe you are salivating or feeling an urge to move the candy around. Those sensations are happening to you here and now. Notice them, and still continue with your commitment of keeping the candy in the center of your mouth. Right here, right now, your mind is likely tossing around lots of judgments, thoughts, and distractions while you are attempting to maintain one commitment: keep the mint centered on your tongue while simply observing what is happening in your inner-world without responding to it.

Any urge to move the candy is happening right now. Notice it. Be aware of this situation: you have committed to keeping the candy in the middle of your mouth, and there are lots of things cropping up that might influence you to stray from your commitment. Notice the thoughts. Accept the feelings. Motivated by the things you care about – your health, family, independence – see if you can stick with this exercise of being right here and right now, so you can use this same skill of situational awareness when you are in the workplace. (Just like the other exercises, you can see Appendix B for resources for obtaining this exercise on audio recording.)

"Here Now" exercises can be tailor-made

You can simply practice being here and now with almost anything you do everyday. Pick a daily activity you usually do privately. Drinking your morning coffee will work just fine as an example. When you drink your coffee, purposefully be present with that experience. Feel the heat of the liquid, notice the flavor of your drink, and observe the coffee aroma in the air. Stick with that experience. Don't read the paper. Don't check your emails. Don't take phone calls. Resist the urge to plan your daily schedule or float off into a daydream. Be here and now with your coffee. If (and when) you do drift off into planning your schedule or daydreaming, simply acknowledge that you were distracted, and then refocus your attention back to the coffee. And when these internal distractions occur again and again, focus and refocus your attention back to your committed action.

You can take several approaches to improving your situational awareness. I know a contractor named Nick who is very committed to practicing his situational awareness exercises, but he struggled because he was always traveling for his job. Nick decided to begin focusing on something he could do no matter where he went. After working a full day of construction, he would go back to his hotel, shower, eat, and then practice his here and now skills. He would find a safe place to go for a walk and simply do that one thing: walk.

Like I suggested in the last chapter, Nick simply committed to paying attention to one thing: the feeling of his boots coming in contact with the sidewalk. That would be his commitment. He was motivated to be better at contacting here and now because he knew it would improve his situational awareness and help keep him safer on the job. Whenever something popped up in his mind and distracted him from paying attention to his footsteps, he would gently refocus himself. If stressful emotions cropped up during his walk, or he got "caught in a thought," he would use his skills of acceptance and noticing to deal with those obstacles to his commitment, and then he would refocus on his walking. If he noticed an obstacle on the sidewalk (a crack, a car entering a driveway, another pedestrian), he would of course acknowledge it's presence, focus on taking correct action, and when it was safe to continue to be mindful of his footsteps, he would resume sharpening his concentration with that exercise.

You can practice situational awareness when you do almost anything. Some people do it when they are washing dishes, lifting weights, or gardening. The action that you choose to be more mindful of isn't really all that important. The really important part of this is that you actually *do* the exercise. Pick an action you can promise yourself you will do everyday. When you do this action, focus on it. Don't multitask, daydream, or get distracted. Practice this type of exercise for 30 seconds a day for a week, and your attention span will increase when doing your chosen task. If so, expand the exercise to 60 seconds.

Build up your attention span, increase your awareness of that situation, and stay in the here and now. I invite you to increase the duration of your mindfulness, so you create a daily practice of paying attention to one thing (your breathing, a mint, your coffee, your footsteps) for a few minutes a day. It is bound to accelerate your ability to stick with your various commitments.

Tough Guys Do This Stuff!

I can remember the first time I taught about increasing situational awareness by doing breathing exercises. The audience was filled with guys who worked in a rough industry, and after the workshop was finished, a man with a shaved head and cauliflower ears approached me to talk. I was partially ready to hear some criticism about how these breathing exercises were just fluff, but what happened next surprised me. The guy said, "I've been doing this mindfulness stuff for years. When I started taking martial arts as a teenager, my sensei not only taught me breathing exercises to help me have more powerful punches and kicks, but he stressed the importance of breathing as a way to stay focused and aware in the middle of a match." Three other audience members gathered around and started talking about how they were trained in very similar methods as martial artists and athletes, not only to improve performance, but concentration as well.

After that conversation, I decided to research other people in the world who say that their performance had been improved by practicing being mindful. I quickly discovered that three of the greatest mixed martial arts champions in the world - Chuck Liddell, Fedor Emelianenko, and Lyoto Machida - do the types of exercises I talk about in this chapter. I figure that if those tough guys can see the merit in the exercise to keep them focused on their jobs, then it really can't be considered a wimpy practice to simply follow your breathing. On the contrary, exercising your ability to focus on what you're doing is a very strong and tough thing to do.

Successful people do this stuff, too!

People in other careers also find merit in doing daily exercises to increase their situational awareness. Phil Jackson was an NBA basketball player, and is arguably the most successful NBA basketball coach of all time. He promotes the idea of learning to be in the here and now in order to promote solid performance. In his book *Sacred Hoops*, Jackson talks about how he taught Michael Jordan and the other Chicago Bulls how to be more situationally aware on the court through exercises similar to the ones discussed in this chapter. Under Jackson's guidance, the Bulls had a historically amazing streak of NBA championships, and Jackson suggests it had to do with their situational awareness.

Other people who have reached the absolute pinnacle of success in their industry admit they practice being in the here and now on a regular basis. Paul McCartney practices being mindful of what he is doing, and *The Beatles* have sold more records than anyone else on the planet. Jerry Seinfeld also includes here and now exercises in his daily routine, and *TV Guide* calls his show "the greatest television show of all time." Many highly successful people can attribute their success to practicing the skill of being in the present moment: Clint Eastwood, Mick Jagger, Elle MacPherson, Albert Einstein, and the list goes on.

People who are skilled in being here and now are surely more likely to succeed with their commitments. If you think about it, suppose UFC Champion Chuck Liddell is in the ring for a title fight.

Would he perform better if he was thinking about what is going on right here and now inside the octagon, or if he was reminiscing about the last time he won the belt? Do you think Michael Jordan scored so many three pointers while he was on the court because he was focused on the game, or because was thinking about where he was going for dinner after the game was over?

Do you think *you* will be better at detecting job hazards if you are distracted, or if your head is in the game? Even further, won't you deal with the hazards you detect more appropriately when you can stay focused on your task? Obviously, you want to be focused while at a dynamic jobsite, and situational awareness is a skill you can build with these exercises. Taken as a whole, being in the here and now helps you *act in the direction of what you care about even in the presence of obstacles.*

Addressing Criticisms of Practicing Situational Awareness
Practicing being in the present moment might be a little "out there," but it is a great skill to have if you want to improve your safety performance. Lots of good ideas sound strange at the beginning. Let me address some concerns you might have.

Is this meditation? Are you asking me to meditate?
I'm not asking anyone to meditate. I'm not asking anyone to do anything. This is simply an invitation to improve your ability to stick with your commitments. I suppose some folks could argue the things in this chapter are related to meditation, but to me, they are related to improving safety. From my perspective, meditation is often linked to spirituality, religion, and enlightenment. If you do that stuff, I think it is wonderful. But that's not what I am advocating here. I am

simply advocating you practice the skill of paying attention so you can broaden your situational awareness.

You don't need to sit on the floor and gaze at a candle while chanting in order to be better at contacting the here and now. (If you choose to expand your situational awareness by doing those things, then good for you.) You can simply practice being in the here and now by doing your daily activities, and making sure you focus and refocus your attention on what you are doing. The more you exercise, the stronger you'll be.

People's minds wander all the time and they don't get hurt!
Your mind can wander quite often and you can get away with it. If you are well-trained in detecting unsafe situations, and then some external-world danger enters into your line of vision, you will likely spring into safe action, even if you were distracted by your own thoughts or emotions. The point I'm trying to make is that over the long haul, you're better off if you can be mindful of what you're doing. If you are prone to think about things other than what you're doing, you increase your risk of an injury.

Increasing your situational awareness skills can be helpful. Saying you don't have to build up your mindfulness skills because you haven't been hurt yet is like saying you don't need to wear steel toe boots because you haven't had anything drop on your foot yet. Building safety commitment requires being able to be more aware of your situation.

How do these exercises help safety?
You might be wondering how exercises about paying attention to a mint, your breathing, or your walking will help you when you are in

the workplace. The idea of practicing these exercises while you're *not at work* is so you can learn to concentrate while you are in a safe, private environment. Your ability will be built up after repeated exercise. Ultimately, you will transfer your situational awareness skill over to the workplace. In times of trouble at work, you will have increased ability to refocus on what you're doing.

Consider the practice of these here and now exercises with this analogy: Many fire departments require firemen to regularly use gym equipment – even during the work shift - because performing specific repetitions of exercises builds up certain job-related muscles. When a fireman's muscles are stronger, he can do his job better. The department doesn't expect firemen to simply become stronger by running into burning buildings, dragging hoses, and lifting people. On the contrary, the department requires using the gym under safe conditions so the fireman is strong enough to do strenuous work in dangerous conditions.

Do I really need to practice?

Sometimes when I suggest using these exercises to build situational awareness, people say, "I'll just make myself be more aware, I don't need to practice!" That is like a doctor telling an overweight person to change his diet and the patient saying, "I'll just make myself skinnier, I don't have to change what I eat." If you always do what you've always done, you'll always get what you always got.

Practicing pays off. If you put forth the effort to improve this skill of situational awareness, you will not only improve your ability to detect and react to danger, but you will also strengthen your commitment to safety.

The Building Block of Your Safety Commitment Plan:

Here, Now

Your commitment plan is strengthened by including reminders to be in the here and now. When completing your Safety Commitment Plan checklist prior to a work task, take a few moments to truly follow through on this item: center your situational awareness on what you are doing, notice your environment, and let go of distractions about the future and the past. Stay in the present moment.

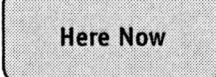 **Here Now**	Center your situational awareness on what you are doing in this work environment. Notice what is happening here and now and let go of distracting thoughts about events not in your present control.

Chapter 8

I Am

The topic of this chapter is important to building safety commitment, and it's a little tricky to explain. Please allow me to start out with a story. Abe, Juanita, and Trevor were veteran workers at their local manufacturing plant. The three friends were given the task of renovating an empty warehouse. During this project, they spent a good portion of their week standing on 20-foot scaffolding while dissembling exhaust fans and ceiling lights.

One day at lunchtime, the three friends prepared to leave the worksite and walk to their favorite restaurant. As they were taking off their PPE near the job box, Trevor snapped his fingers and said, "Hey, I promised the boss we'd finish taking out all the lighting in the southwest corner of the warehouse by noon today. I just realized there's one more ballast to take care of and then we'll be done. Why don't you guys go to the restaurant without me? Order me my usual lunch, and I'll go back up the scaffolding and finish up. I'll catch up with you both soon."

The other two workers agreed and started heading out the door. Juanita turned around and noticed that Trevor did not put his fall-protection harness back on. As Trevor climbed the scaffold, Juanita called out: "Hey, you forgot your harness!"

"Don't worry," Trevor yelled, "I am safe."
Trevor's internal-world gave him an inaccurate description of himself. He perceived his self as "safe." But the external-world showed him his perception of his self was wrong! He fell from the scaffold and broke both of his legs and his left arm.

Everybody has a self-perception. We describe our self in many different ways. We can give descriptions of our physical features, our personality, and the roles we play in our lives. We can talk about how we relate to people, where we grew up, and what we enjoy eating for dinner. We can talk about our self in many different ways. Our ability to describe our self is an important skill.

"Know thyself" is a famous phrase from philosophy, and it has been passed down for centuries because *it is really good advice.* Knowing yourself is important to living well. But as we move through this chapter, you'll recognize that some of the things we say about our self aren't always true or helpful when it comes to safety. The aim of this chapter is to help you realize that you should be careful how you talk about your self.

The "I Am" Exercise

I'd like to invite you to take a look at your self, and I'm not asking you to see your reflection in a mirror. Instead, I'd like to ask you to reflect on who you are in your mind, and then write it down on in the box below. Think of this exercise as the easiest fill-in-the-blank quiz in the whole world.

I am...

I am _____.

I am _____.

I am _____.

I am _____.

I am _____.

Please don't move forward without doing this. If you don't want to write in the box, use scrap paper. If you don't have a pen, then please at least do this exercise as diligently as possible in your head. There are five blanks you have to fill in. Please come up with five solid answers.

Once you've completed the exercise, let's move forward. I imagine it was rather easy to come up with these answers. You know your self pretty well, and it's fairly simple to describe who you are and what kinds of characteristics you have.

Now, I'd like to you to try something with your list of descriptors. Try to continue with this exercise even if it feels strange at first. I want you to take the first answer you wrote in the box and just cross it out. Now let me ask you a question: Are you still you?

I understand this might be an odd exercise, but the question stands: if you couldn't describe yourself with the words in that first answer, would you still exist?

When I do this exercise in workshops, I sometimes get a sense that people might not feel comfortable crossing out a descriptor, but most people will still admit they do exist, even without that one descriptor. The fact is, you are "still you" if you cross out that word or phrase from the first line.

What if you were to cross out the second answer? If you couldn't describe yourself that way, would you still exist? I imagine the answer remains "Yes." (Some people insist they wouldn't exist without certain descriptors; we'll discuss that issue later in the chapter.)

Now cross out the last three descriptors you wrote. Without those descriptions, see if you can come to terms with the idea that *you still exist*. And now I have a new question for you: What words remain in the box? If you think that you crossed out all the words in the box, check again. There are some pretty profound words in the box that still describe you. Those two words are: "I am."

The two words remaining in the box form a complete sentence, and "I am" is one of the few things that you can say about yourself that will never change. You can never really cross that statement out. You exist. *You are.* When you say that from your own point of view, you can declare, "I am." You can know that will always be true for the rest of your life.

When you boil it all down: you exist. You have your very own unique point of view, and you have experienced everything that has ever happened to you in your life. However, most descriptors that come after "I am…" do not really describe you in total, and they aren't always true. Everything else comes after "I am." If this seems like an unusual safety topic, it is! But stick with this chapter because we're talking about how your own self-descriptors and your own self-concept can become obstacles to safety. Let me illustrate the purpose of the "I am" exercise with an example.

Using my answers as an example

When I completed this exercise, here is what I wrote:

I am...

I am *sitting* .

I am *writing this book* .

I am *happy* .

I am *a safety consultant* .

I am *six foot, three inches tall* .

Those five answers are simple descriptions of how I was identifying myself at the time. You can look at those answers as part of my "I"dentity because those answers describe a little bit about me at the time of writing. It's important to understand that these five phrases certainly do not describe me completely, and they also have varying degrees of accuracy. The truth of these descriptions changes because so much about me changes.

Take the first description of myself: "sitting." This is true: I am sitting right here and right now as I write this book. But my *posture* changes all the time. Sometimes I am standing, kneeling, or lying down. By the time you read this, I might not actually be sitting. The same goes for my second answer: "writing this book." When I wrote that as my answer, it was true. However, since you are reading this book right now, I am obviously not writing it anymore. I am doing something else. Right now, as you read this, I could be shoveling snow, mowing my lawn, or playing basketball with my two children. (I sure hope it's the third one!) My *behavior* is constantly changing.

The third descriptor is also something that changes. I am "happy" as I work on writing this book right now, but I'll likely be angry if my editor gives me bad feedback. *Emotions* change all the time.

My profession as "a safety consultant" might seem like it is something a bit more permanent than my posture, behavior, and emotions… but it will change, too. I will hopefully retire and then I will no longer be a safety consultant. The roles we play in our lives vary. You may wear many hats, but you still only have one head.

My height seems relatively enduring. I mean, I've been "six foot, three inches tall" for over two decades, and I hope to be that tall for another 20 years. But people shrink as they get older, and our bodies change over time. Our body is in flux most of the time.

So we can describe ourselves lots of ways: our posture, behavior, emotions, profession, and physical characteristics. We can also describe our memories, the roles we play, the thoughts we have, and the sensations we feel. We can use lots of words to describe our self. There are many descriptors that can come after "I am," but they aren't necessarily permanent and they aren't always true.

Now, you may be asking yourself, "How does this relate to safety!?" Before I link this to building safety commitment, I want to tie up one loose end.

I'm not me if I crossed out one of my answers!

If you are having trouble with the "I am" exercise results because you don't feel like you still exist if you cross out one of the answers, I have a few things for you to think about. First, the main point to this exercise is to show that an overwhelming number of self-descriptions are either temporary or incomplete. If you came up with one that you feel strongly about, let's not let that get in the way of the main point. I simply want you to experience the idea that *some* of the things we say to ourselves just aren't always true or helpful. If there is something in this exercise you feel strongly about not crossing out, leave it… as long as it doesn't get in the way of your commitment to safety!

Also, I know that if one of your answers to the exercise was "alive" or "living," then the exercise doesn't work so well with that answer. The reason it doesn't work is because "I am" and "alive" are almost the same thing. If you cross out "alive" then you can't really say, "I am." It just turns the exercise into a strange contradiction or a useless word game. If you answered that way, just skip that example.

You might have used some other descriptors that are permanent. For example, I could have put: I am male, I am Irish, I am from New York. Those things will never change. But if you chose a more "permanent" description, keep in mind it still does not describe you *in total*. When I say, I'm an Irish guy from New York, that is just scratching the surface. Self-descriptors can't *completely* describe a person.

You might have written, "I am…"

…a parent.

…interested in safety.

…eternally in love with my spouse.

…made by my Creator.

…caring.

And I imagine if you wrote something along these lines, you are reluctant to cross those things out. You might argue that without these descriptions, you wouldn't be who you are. And I would suggest those descriptions do not describe you in total, and they aren't always your focus. This exercise is not about debating what is important to you. The exercise is simply an attempt to loosen up the language you use to describe who you are just in case it gets in the way of your commitments.

Going back to the story at the beginning of this chapter, can you recall the last thing Trevor said to Juanita before he climbed the scaffold without wearing fall protection? Trevor believed something about himself that wasn't helpful. He said: "I am safe." In his dynamic work environment, it was unhelpful to believe that kind of description. Those are just words that come from his internal-world. They didn't actually describe *the reality*. That descriptor was an internal-world obstacle to safety and led to dire consequences.

Once again, the most important point is for you to understand this critical idea: when you say something about yourself, it might not be complete, true, or helpful. And here is why I am making that point: Which do you think is more helpful for creating optimum safety:

a) *believing the words* you say about yourself are true, or

b) *acting in the direction of what is important even in the presence of obstacles?*

Obviously the answer is "b." Keeping safety commitments is much more important than believing your descriptions about yourself. And here's the problem: you have probably said things about your self that became barriers to safety! Sometimes the things you say to describe yourself *create obstacles to your commitments.* Some of our beliefs about ourselves are delusions.

Alcoholics sometimes say "I am able to stop whenever I want," and abused women have been known to say, "I am fine. I am sure this relationship is the best thing for me." At worksites, some young adults carry around a rough-and-tumble attitude of "I'm able to take it," or "I'm bulletproof." Lots of workers seem to complacently take on the personal belief "It won't happen to me."

Our self-descriptors can distract or mislead us into a false sense of security. Using caution with what comes after "I am…" is the point of this chapter. The two-word sentence "I am." will be true for you for the rest of your life. As far as the other things you say about yourself, hold them loosely because they might not always be helpful to you.

Believing Unhelpful Self-Descriptors is a Safety Obstacle

Throughout the last four chapters, we discussed four internal-world safety obstacles: lack of values-based motivation, stressful emotions, getting "caught in a thought," and getting distracted by a wandering mind. A fifth obstacle to optimum safety is *believing unhelpful descriptions about your self.* Trevor bought into a self-descriptor that influenced him to skip taking an important safety step, and then he experienced a tragic incident. "I am safe" might be a nice thing to *believe*, but it's even nicer to *engage in actions* that help ensure that description. I have heard several other unsafe and unhelpful self-descriptors from people over the years:

- "I am almost positive we did the lockout/ tagout."
- "I've been trained. I don't need anymore talk about it."
- "I am a veteran here. You can't teach me any new tricks."
- "I'm gonna be fine. It's just this one time!"

These four phrases are examples of things you can say about yourself that aren't very helpful, and we can address each of these problematic self-descriptors by going in order:

- Telling yourself that you're "almost positive" is no substitute for the action of going and checking the LOTO box.
- Knowing how to do a job task does not mean elements of the work have not changed. There could be new things to consider when working. Believing you don't need further instruction is the kind of rigid thinking that puts people at risk for new hazards.

- It's great to have pride in being a veteran worker, but have you succeeded in your long career because you were lucky, or because you were continually vigilant for dangers? It is unhelpful to think that years of experience make you immune to unexpected incidents. All precautions should be taken, including getting new training.

- Saying "I'm going to be fine" is just another "I am" descriptor that is temporary and potentially inaccurate. You might be fine when you start doing the unsafe thing, but safety is not just a part of *you*… it's related to the external-world, too. All the positive thoughts and beliefs are no match for a 20-foot fall from a scaffold… just ask Trevor.

The "take-home" point is that you should be very careful when you describe yourself because some thoughts and beliefs might impede safety. Lots of things we say about ourselves aren't always true, and it's much better to follow through on committed actions that ensure safety, rather than just believe the words "I am safe."

Isn't Having Positive Beliefs about Myself Good?

Healthy people typically have positive thoughts about themselves. I am not criticizing positive thinking. If you want to believe positive things about yourself, I'm not suggesting you stop. Just make sure the thoughts are helpful to you! If you say "I am safe," and that statement reminds to wear PPE, then that positive thought was helpful. On the other hand, it's not helpful if you say, "I am safe" and use that phrase as an excuse to *not wear* your PPE.

As I suggested in Chapter 3, it's important that your internal-world experiences help support your safety commitment. Having false or unhelpful beliefs about your self is an obstacle to safety. Instead of telling yourself "I am safe," try telling yourself, "I am here now, accepting the way I feel, noticing my thoughts, while doing what I care about." Now that is an "I am" descriptor that will help you keep your commitment to safety!

The Building Block of Your Safety Commitment Plan:
I Am

I know this chapter talks about a unique topic for a safety book, but I believe it's important for everyone to treat their own self-perceptions with caution. Every one of us is simply a human being. We can get hurt and some people get fatally injured. We need to be careful about thinking we're special, bulletproof, or safe. The words you tell yourself – the things you believe about yourself – could make you drop your guard at the wrong time.

While completing your Safety Commitment Plan checklist, recognize the fact that you are susceptible to internal-world distractions and external-world dangers. If you're being hooked by any unhelpful self-descriptions or any self-perceptions that make you drop your guard, skip steps, or become complacent, see if you can notice those thoughts and beliefs are not helpful, let them go, and simply recognize the only true description: "I am."

I Am	Notice if you are being influenced by any unhelpful self-descriptions. Let go of anything unhelpful that you are believing about yourself.

The Third Step:

Planning Committed Actions

Chapter 9

Doing

The previous five chapters focused on what you can do to remove internal-world obstacles hindering safety commitment. This chapter is different from those five chapters because it bridges the internal-world and the external-world by focusing on the sixth domain on the Safety Commitment model: Doing. As you read this chapter, it will likely feel different than the previous ones because it is directed about how you design your work tasks for optimum safety, rather than how you can improve your internal-world for safety.

This chapter is about actually interacting with the external-world, and *doing* something about your own safety. Earlier in this book, we said a commitment is more than just talking about action; it is *doing* action. A commitment is "*acting* in the direction of what you care about even in the presence of obstacles." Safety commitments require truly measurable actions in the external-world. All companies need their employees to engage in certain behaviors in order to ensure safety in the workplace. Executing a To-Do list targeting specific safety actions in the external-world will help you move forward on your safety goals and Safety Commitment Plan.

Critical Principles of Making a To-Do List

There are several reasons we want to include a To-Do list for External-world Safety Actions in your commitment plan. Most importantly, the definition of commitment requires "acting," so a commitment plan should define what to take action on. You cannot just say, "I will commit to being safe" because that is too general and not easily measurable.

Instead, you can develop your To-Do list by taking inventory of all the steps needed to successfully complete a job task, and then document those steps so they can be followed in a checklist fashion. Once you know what task you are committing to do, you will benefit from dividing the task into smaller objectives and pinpointing each of the steps you will be taking. By pinpointing, I mean you should precisely describe each action step in a sharp, unmistakable manner. A really good pinpoint describes either your behavior or the result of your behaviors. Pinpointed action steps are best when they can be observed, measured, and recorded.

Doing	To-Do List for External-World Safety Actions	

Once each of the steps for proper execution of the task has been pinpointed and put in sequence, you can use your To-Do list as if it were a checklist. Create a column in your list where you can tally whether or not a step has been accomplished. I'm sure you have noticed that each building block to our Safety Commitment Plan from the last five chapters included a box on the right-hand side so they can be "checked" when completed. Your safety commitment will be greatly supported if you use a checklist to keep you focused on what you should be doing step-by-step and in the moment.

Making a To-Do list is actually a very powerful way to make sure you stay on track with your safety commitment. However, some people might believe To-Do lists seem like a wishy-washy attempt at ensuring safety in a dynamic work environment. Workers tend to be resistant to following checklists. In Atul Gawande's famous book *The Checklist Manifesto*,[17] he describes how impactful checklists can be for promoting success and safety in many professions. When he highlights the medical profession, he notes that even when doctors and nurses are trained in the effectiveness of checklists for keeping workers on task, they remain skeptical about adopting them into the work stream. However, he also noted that when those same medical workers were asked if they would want their surgeon to use a checklist during their own surgery, 93 per cent said "yes!"

That goes to show people see the utility of checklists but are reluctant to actually use them. Strongly consider making a commitment to using the checklist function of your commitment plan. After all, if you are going to commit to something, you should commit to safety plans and methods that have been shown to work effectively!

When developing your own To-Do list, you benefit from separating a larger task into smaller, discrete steps and checking them off one after the other for a few reasons:

1) You can monitor what is getting accomplished,

2) You keep complex tasks flowing in sequence,

3) You can make sure that nothing important is skipped, and

4) If you get interrupted you can efficiently begin where you left off.

Many job sectors (construction, aviation, medical) have integrated the use of checklists to help workers follow a uniform process, and the importance of checklists for improving safety have been documented.[18,19] Creating a checklist can reduce unwanted variability in your actions and assist in prioritizing your work behavior. Checklists also help you remember where you are in a complex work process and can improve delegation of tasks and teamwork.

Of course, a checklist has limited power because it only includes known hazards, so workers and leaders must maintain their situational awareness for new problems, and get on-going training about potential sources of incidents in their industry. An additional consideration about the use of checklists is the chance you could deviate from actually following it! You can prevent deviation from happening in a number of ways.

First, you can put this book to work for you: while following your To-Do list, notice any distracting and unhelpful thoughts, stay in the present moment, and accept the fact that even if you don't feel like following a checklist, you can stay committed anyway.

Second, you can also use your personal values as motivators that influence you to stay on track with your checklist. Beyond those tools, good To-Do lists include *performance management* techniques that make the commitment plan function like a contract. I will discuss performance management later in this chapter.

Because there are thousands of different occupations, this book cannot outline all the To-Do lists workers need to follow to maintain safety on the job. Many jobs are unique, and most To-Do lists have to be specifically targeted to *your* particular work environment. Let's take a look at the preliminary steps for developing a To-Do list.

Creating the To-Do List with a Job Hazard Analysis

Jobs for front-line workers have become significantly complex over the years, and the dedicated use of To-Do lists will help simplify tasks and can reduce the number of errors made during the workday. When a group of employees are doing identical tasks, a job-specific To-Do list helps them literally be "on the same page" when it comes to work and safety methods.

A To-Do list should describe the best practices for completing a job. ("Best practices" describe the most efficient and safest way to complete the job as reviewed by professionals in that industry). Of course, best practices evolve in the face of new discoveries and improved equipment, so when you're developing a To-Do list, make sure you're educated in the most effective ways to maintain safety in that situation.

An excellent way to create a To-Do checklist is to first perform a job hazard analysis (JHA). In OSHA's definition, a job hazard analysis is a method for identifying safety problems before they occur and "focuses on the relationship between the worker, the task, the tools, and the work environment."[20] OSHA strongly suggests organizations add JHAs to the company safety plan in order to prevent injuries, establish proper job procedures, and ensure all workers are trained properly.

Typical JHAs describe the job location, a task description, a hazard description, and steps for controlling the hazards. JHAs give you discrete steps to follow, and can become a blueprint for your To-Do list. Let's take a look at an example on the next page of a JHA used at a chemical plant. As you can see, there's more information in this JHA than you need to develop a To-Do list, but it makes sense to introduce you to the entire JHA example so you are exposed to other best practices used for establishing a safe environment. This particular JHA describes the location and description of the task, as well as the potential hazards to prepare for during the job.

Many JHAs include the name of the safety analyst just in case there are questions about the procedures. The date of the analysis is important because it makes sense to review and update the JHA after a renovation, introduction of new tools, or other events that might make old safety processes obsolete. JHAs should include the Sequence of Events for the task (ex. "Enter the tank," "Cleaning the tank"), and the procedures the worker should follow to prevent incidents and injuries during each step.

Job Location: Chemical Tank	Analyst: Angelica Grindle	Date: 05/15/2013

Task Description: Worker enters an empty tank in order to scrub the lining to prevent future contamination of product chemicals.

Hazard Description: This task has a potential to expose a worker to the gas or liquid in the tank, trip and fall risks, ladder risks, and the risk of handling heavy equipment. While cleaning the tank there is risk of reaction to chemicals and contaminants released into the air from scrubbing.

Sequence of Events	Hazard Control Procedure	✔
Enter the tank	Wear personal protective equipment for tank conditions (NIOSH Doc.#80-406; OSHA CFR 1910.134)	
	Ensure another worker is outside the tank doing "hole watch," and is able to instruct and guide the worker entering the tank, and can help lift the worker from the tank in an emergency.	
	Put air quality monitor hose in place	
	Ensure all cleaning tools are present and lowered into tank with transport bucket	
Cleaning tank	Provide lighting for the tank (Class I, Div. I)	
	Provide exhaust ventilation	
	Provide air supply to interior of tank	
	Delegate a person to be on "hole watch" and be alert to air quality monitors	
	Set 15 minute timer to ensure rest periods	
	Provide means of communication from worker to hole watch, and hole watch to emergency team	

The major section of this JHA example can easily be incorporated as a checklist into your commitment plan's To-Do list. All of the steps in the Hazard Control Procedure are specific enough for trained employees to follow, and each step pinpoints actions workers can execute. Each line item begins with an action verb (ex. wear, put, ensure, etc.) and directs behavior in a sequential manner in order to reduce the likelihood of a dangerous incident while scrubbing an empty chemical tank. The information from your company's JHAs can be easily transported into the To-Do list in your commitment plan.

	To-Do List for External-World Safety Actions	
Doing	Wear PPE	
	Ensure a trained and prepared hole watch is on duty	
	Put the sniffer in place and ensure it is working	
	Have all the cleaning tools you need in the worksite etc.	

When you follow a To-Do checklist you are staying on track with your commitment by *doing* safety. You have a specific agenda for *acting* in the direction of what you care about even in the presence of obstacles. When you begin the workday saying: "I am here now, accepting the way I feel, noticing my thoughts, while *doing* what I care about" …you should be *doing* the To-Do checklist steps you gathered from a JHA.

Making your own JHAs

If your company has already developed JHAs for relevant job tasks, they can serve as excellent templates for your own To-Do list. If JHAs are not in place yet, strongly consider beginning this important component to workplace safety. OSHA guidelines suggest developing JHAs by doing the following five steps:

1) Involve workers – Frontline employees have a first-hand perspective on the duties and hazards related to each task. Worker involvement is invaluable, not only because it helps make sure there are no oversights, but also because teamwork can help build their personal investment and commitment to the JHA project.

2) Review the incident history – When compiling a JHA, your company's near miss and injury reports will give you strong indicators of the risks and dangers in the workplace, and how they might be avoided in the future.

3) Discuss a preliminary job review – Talk about each job task with your team. If possible, form a committee that brainstorms about the work tasks, and develop a list of issues for workers to consider while completing the task. The team should also attempt to come up with solutions for safety problems.

4) Prioritize targets for improvement – Ensure the team spotlights any clear and present problems at the worksite. After you have completed your job review, fix the concerns that have the most severe consequence or greatest risk. Work should not begin until the problems have been appropriately addressed.

5) Outline the important steps for each task - Observe a worker in action, and if possible, have a dialogue about the sequence of actions involved, and the potential risky

events that can arise during the task completion. If you notice hazards *while* you're outlining the steps, stop the JHA process and fix the problem immediately. OSHA states "nearly every job can be broken down into… steps," and suggests finding a balance between describing the important duties without being bogged down in meticulous details. That advice also goes for building a To-Do list.

When I work with larger companies, I suggest creating a steering committee of experienced workers and safety champions who will review the latest instructions for doing a particular job safely. The steering committee is then instructed to create a document that outlines how to do that particular job, and to use those directions as a framework for job training and the outline for a proper To-Do checklist. A steering committee is especially important when a job task is unique and there are no industry standards for safety.

If you are not able to benefit from forming a committee, then you're invited to do your own research into creating a commitment plan. Documents explaining the best safety practices are available from reliable resources on the Internet. Attending safety workshops and webinars, and reading the professional articles for your particular industry will help you learn what you should be doing in order to upgrade your safety on the job.

Learning how to complete a professional JHA is beyond the scope of this book, but there are many good resources that give you instructions on how to begin the process. The book *Job Hazard*

Analysis: A Guide for Voluntary Compliance and Beyond [21] offers a solid blueprint for building an effective JHA suited for your organization. When you have JHAs in place, they will help you create a To-Do list that you can incorporate into your own Safety Commitment Plan.

Side bar

In addition to helping improve safety, To-Do lists can also help people keep their commitment to improving their leadership actions. Leaders can easily get distracted from their main focus – leading people – because they are often drawn into dealing with paperwork, participating in meetings, and "putting out fires." Those other duties are important, but a leader shouldn't get so caught up in those tasks that he or she misses opportunities to have a direct impact on workplace safety. Using a To-Do checklist that includes specific leadership actions, such as getting out the worksite, observing workers, and giving feedback helps improve the impact of safety leadership.

All the principles discussed in this book can be applied to improving leadership. In my Commitment Based Leadership (CBL) trainings, I invite leaders to clarify their own values in order to help them envision an admirable direction for their company. We weave together personal values with the company's values in order to accelerate their motivation to stick to their commitment plan. I also use the other concepts in this book to help leaders optimize their skills with greater situational awareness, and improve their abilities for dealing with stress and distractions. Leaders are more effective when they learn the skills we are talking about in this book, and when they work on building leadership commitment.

How to Keep Doing What You Are Doing

There is another piece of the puzzle related to Doing safety in the external-world. To maintain safety actions over the long haul, not only do you need to remove internal obstacles, increase personal motivations, and develop a To-Do list, you also need to have a *performance management* plan in place.

Performance management is the systematic use of behavioral science in order to improve work behaviors. When you alter your environment so that you are well-prepared to do a good job, you are doing performance management. More importantly, when you alter your environment so your work leads to good consequences, you are really doing solid performance management. I consider integrating performance management as part of the Doing portion of the Safety Commitment Model because you are engaging in important actions to arrange the work environment to sustain your commitment.

Performance management is a whole wing of the applied behavioral sciences, so it is impossible to teach the topic in a single chapter, but I'd like to introduce four core concepts you should be doing in order to accelerate safety commitment. Your Safety Commitment Plan will be supported by 1) posting signs and signals to remind you of what you will be doing on a regular basis, 2) publicly announcing what you will be doing for your commitment, 3) forming an accountability partnership with a peer or supervisor, and perhaps most importantly 4) building in a system of rewards for meeting and exceeding your goals.

Signs and signals reminding you about your commitment plan. When possible, keep your written commitment plan where you can regularly see it. Take steps to make sure you get reminded of your safety commitment by placing signs in relevant areas for the task. Making signs and tags for your toolbox can serve as a cue to stick with your commitment. Safety signage *does not* cause you to act safely, and sometimes signs lose effectiveness after being in place for a while. Signs can contribute to helping us focus on doing the safe actions, and arranging them to have an impact on your behavior should be part of your Safety Commitment Plan.

Publicly announce your commitment. Declare to a group of people you will be focusing on your own behavior change. When you do this, your relationships with those people will help you become more committed by showing them what you are accomplishing. To use an example outside of safety, suppose a woman wants to lose weight. Do you think it will be more likely she will diet and exercise if she simply *thinks* about this plan, or announces the plan to her close friends?

Surely, some of her friends will be there to give her positive feedback when she is moving toward her goal, and support her when she is struggling. She might even have people in her social circle who express doubt in her ability to lose weight and then proving them wrong also motivates her to stick with the commitment. Either way, announcing a commitment to other people has greater impact on whether or not you will do it when compared to simply keeping your commitment a secret.

The same logic applies in safety. Tell your friends and work partners about your commitment to improve safety-related behaviors. You can even tell your supervisors if you are comfortable with them giving you more feedback on the job. There is a good amount of research that shows if workers are observed on the job and given feedback about their safe actions, the safety on the site improves significantly.[22] Asking for feedback will help you start to build social support for your commitment to "do safety" and you can supercharge this process by enlisting an accountability partner.

Form an accountability partnership with a peer or supervisor. You can choose not only to announce your commitment, but also to go one step further by having a work partner reliably and purposefully checkup on your progress with the Safety Commitment Plan. The frequent checkups and conversations about what you are doing will help increase your committed actions. It works doubly well if you do peer observations on each other. If you want to significantly enhance your commitments though, you should have an accountability partner assist you in building in consequences for your actions.

Build a system of rewards for your behavior. Unquestionably, the greatest influence on maintaining behavior over the long term is making sure your actions are followed by consequences influencing you to do that behavior again. Let's face it: people do things to get rewarding consequences and avoid punishing consequences.

For an example outside of the safety world: if the consequence of you turning on a radio station is that you hear your favorite songs, you're more likely to turn on that station again in the future. If you eat a particular food and enjoy the way it tastes, you are more likely to eat that food again. On the other hand, if you tune into a radio station that plays music you dislike, you avoid that channel in the future. If you eat food that tastes nasty, you are unlikely to eat it again. Generally speaking, good consequences maintain behavior and bad consequences reduce it.

The same principles work with your safety behaviors. Consequences that follow your actions play a major role in the maintenance of safety plans, as well. Perhaps the most important thing you can do to maintain a solid safety process is build-in good consequences for meeting certain goals related to the commitment plan. You can develop a personal self-improvement plan using consequences tailor-made to improve your own behavior, and it's also possible to devise a company-wide BBS process that uses positive consequences to influence an entire workforce to commit to the safety commitment plans.

If you want to learn more about developing an effective company-wide safety process, *The Values-Based Safety Process* [23] written by my colleague Terry McSween gives you the details for an evidence-based approach of applying behavioral techniques to improve safety processes in the workplace. In brief, scientifically validated safety processes encourage companies to 1) develop To-Do lists for job tasks, 2) create a method for observing the workers' actions, 3) develop measurement procedures for the observations, 4) train observers how to give helpful feedback after observations, 5) chart the progress of how often the workers meet the expectations of the safety process, and 6) plan how to deliver positive consequences to maintain improvements in the safety process measures. When the workers demonstrate the ability to meet their goals, positive rewards are presented to help the workers want to continue to work safely. When these implementations are executed properly, they have an excellent impact on injury rates and profitability. In addition, the safety culture and employee morale are improved.

Performance Management Contract	
Signs and signals are in place	
Publicly announced safety commitment	
Accountability partner: _____ is aware of my commitments	
Describe incentives and performance criteria related to safety process:	

My colleagues and I have implemented effective performance management processes around the world, and our clients have demonstrated remarkable improvements with safety performance and safety culture. Injuries were decreased, financial costs were diverted, and productivity increased. The implementation of company-wide commitment plans improved the company's bottom line while creating a healthier work environment for the employees.

You should use the checklist in the Safety Commitment Plan to make sure you're setting up the correct reminders for your safe actions: signs, public announcements, and an accountability partner. Feel free to tailor these preparations in a way that will be helpful to your specific situation. Extenuating circumstances will require you to think flexibly about making it work for you and your worksite, but make an effort to integrate these ideas and then check each item off after completion. In addition, make sure to briefly describe the performance management contract consequences at the bottom of your commitment plan as a reminder of the positive outcomes for doing the right thing, and in order to increase external motivation for the commitment plan.

The Building Block of Your Safety Commitment Plan:
Doing

Behavioral science has helped develop applications for altering the external-world in a way to help people continue doing certain actions. If you care about safety, then you must investigate what actions you need to be *doing* to ensure you are safe and then build a *performance management* system around those actions to make sure those actions continue. Once you have honed what you should be doing, incorporate the other topics from this book to help you keep your commitment moving towards what you care about (values-based motivation), and to stay dedicated even when obstacles arise (using acceptance, noticing, a clear sense of your self, and situational awareness).

Remember, a commitment is "acting in the direction of what you care about even in the presence of obstacles." Doing the To-Do list is "where the rubber hits the road" in a commitment. To explain that analogy a bit further, it takes only a small area of contact between the tires and the asphalt to move a whole automobile forward, but that small space is critical. You can have a beautiful classic automobile with a fine-tuned engine, clean interior and beautiful paint job, but you need the rubber of those tires to meet the road if you want to go anywhere. Those four tires are the most important part of the car if you are interested in building up traction and moving forward.

The same goes for creating a strong commitment: you can have fine-tuned motivation and excellent skills at dealing with your internal obstacles, but if you aren't *acting* on a solidly constructed

To-Do plan, you aren't getting any traction on what really matters. So if you'd like to see growth and positive movement in the direction of safety, use this To-Do list template as a building block to your commitment to safety, and be explicit in itemizing the important behaviors you'd like to see completed in order to ensure safety. And to ensure you maintain those actions, make sure the antecedents and consequences are built into your Safety Commitment Plan.

Let's look at a brief example to illustrate how to apply these two building blocks for a Safety Commitment Plan. Tyler is a rookie machine operator at a mid-size manufacturing plant. During training he's taught the important actions required for his safety and performance while on the line, and these steps are reviewed at the morning toolbox talk on a regular basis. One day, Tyler reviews the safety steps in his mind, checks them with his supervisor, and fills in the To-Do list on his Safety Commitment Plan.

Doing	To-Do List for External-World Safety Actions	
	Wear proper gloves, eyewear, and ear plugs	
	Lift the material bundles with hoist	
	Remove the bundling straps with snips	
	Load the manufacturing equipment with materials only when the green light is on	
	Discard the bundling strap in recycler	

It's also important for Tyler to make sure he's aware of the antecedents and consequences related to his performance management. Prior to working, he makes sure that he's doing the work he's been trained to do, and there are reminders channeling his actions toward safe performance.

During the toolbox talk, the foreman asks Tyler what he's committed to doing, and Tyler announces his planned commitment to adhere to PPE standards and to make sure he treats the materials and equipment with the respect and care required. His coworker Jonathan does the same job on a different machine, so they promise to keep an eye on each other. They affirm to each other they'll give proper feedback, do it in a constructive manner, and won't hold back if they see a problem. They have become accountability partners, so Tyler writes Jonathan's name in his Safety Commitment Plan.

Tyler's company has implemented a BBS process where peers occasionally come to a worksite, ask permission to do an observation, and then record if the employee was working safely. If required, feedback is given in a helpful manner. The data are collected and if 80 percent of the observations meet the safety criteria, the safety committee holds a celebration event: a Friday Fish Fry for the shift crew. Tyler writes these details in his Safety Commitment Plan as a reminder of how the external-world influences on his safety actions.

Performance Management Contract	
Signs and signals are in place	✔
Publicly announced safety commitment	✔
Accountability partner: _Jonathan_ is aware of my commitments	✔
Describe incentives and performance criteria related to safety process:	
If 80% of the observations meet safety criteria by end of the month, everyone on the shift earns the Friday Fish Fry for lunch.	

These two building blocks of safety – the To-Do list and the performance management contract – are crucial aspects for bridging internal-world skills with external-world actions. The two building blocks fill in the final "Doing" portion of the Safety Commitment Model. Now that the model is complete, let's turn to the next chapter to bring it all together.

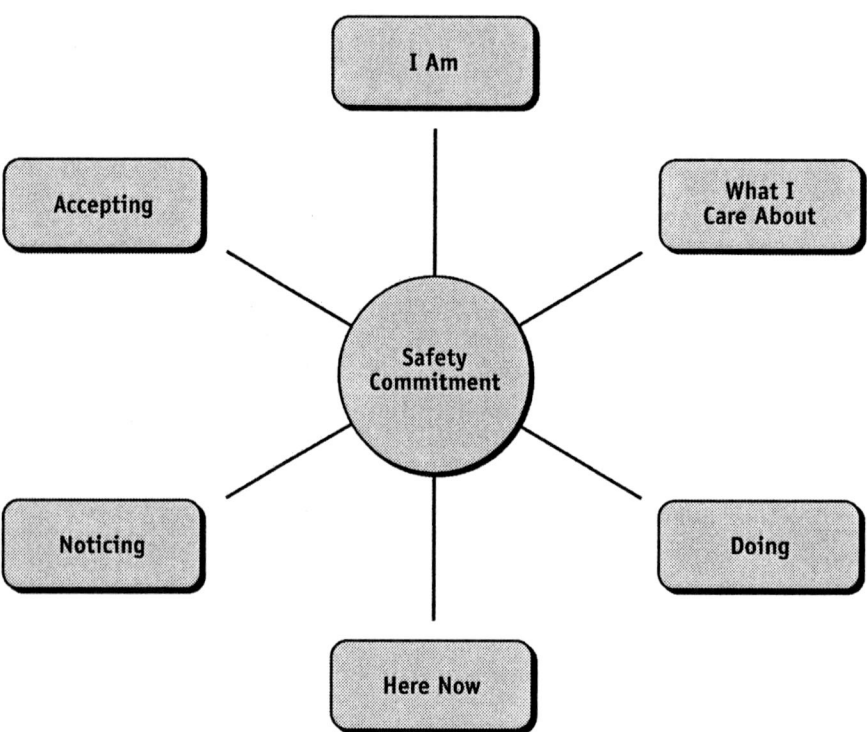

Bringing It All Together

Chapter 10

Using Your Safety Commitment Plan

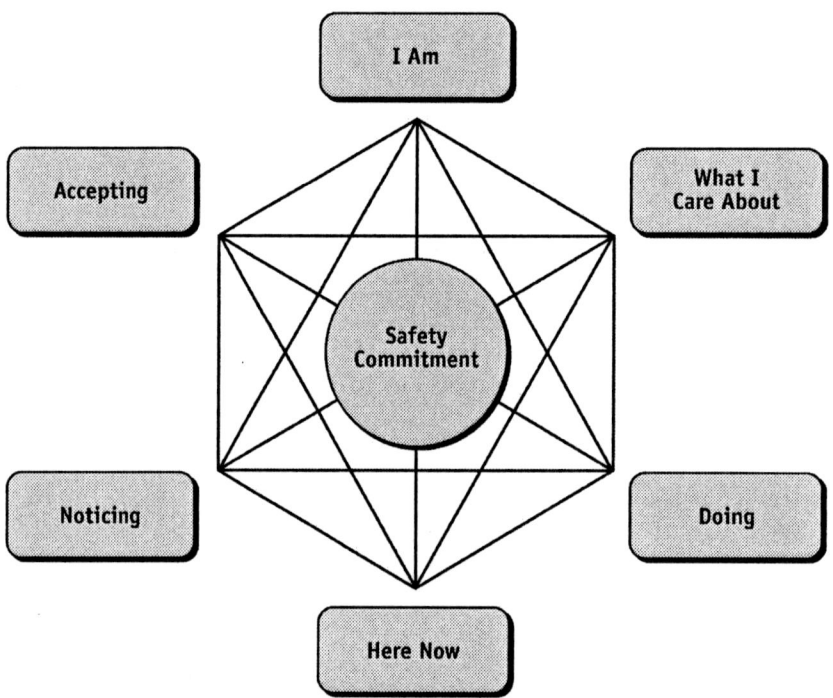

The central theme of this book is building safety commitment, and you can see it's at the heart of the six topics you've been learning about in the last several chapters. When you look closely at the diagram, you will see the six concepts we covered in this book create a simple cornerstone phrase aimed to improve safety commitment:

"I am here now, accepting the way I feel,

noticing my thoughts, while *doing what I care about."*

Building skills in the six topics related to that phrase will enrich your safety commitment by helping you develop a healthy self-concept, become more aware of your present situation, learn how to deal with unhelpful thoughts and emotions, and clarify your values to increase your motivation for acting safely. The previous six chapters gave you the building blocks you need to follow an effective commitment plan. We've been talking about these six areas as if they were separate, but they all draw on each other for strength. Each of the six skills can be used separately to boost your ability to deal with internal-world obstacles and help you execute external-world safety action, but there is added synergy coming from the combined use of the components of the Safety Commitment Model.

When you have answers for "What I Care About," then not only are you more motivated for safety, but you are also more motivated to show willingness to accept tough emotions.

When you practice your ability to be "Here Now," you get better at simply "Noticing" thoughts and "Accepting" emotions.

When you are willing to let go of unhelpful descriptions about your self (ex. "I am a veteran so I don't need more training!), and come in better contact with the "I Am" experience, you are more likely to start "Doing" new things like going to safety training workshops.

The skill of "Accepting" helps you with "Doing." The skill of "Noticing" helps you focus on "What I Care About," and so on. I could continue to illuminate all of the ways each of these skills strengthens the others, but it is more important for you to realize that you can bring all of them together to strengthen your safety commitment. Let's bring each of the building blocks together.

The Safety Commitment Plan Worksheet introduced in Chapter 3 was simply a blank framework of the concepts discussed throughout the book. At the end of each chapter that followed, we filled in one piece of the outline with a summary of the skills from that chapter that you can apply to building stronger safety commitment. Let's review each component of our cornerstone statement as it pertains to the Safety Commitment Plan Worksheet.

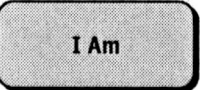

I Am

"Notice if you are being influenced by any unhelpful self-descriptions. Let go of anything unhelpful that you are believing about yourself."

Your commitments are more likely to be carried out when you are able to let go of any unhelpful beliefs you have about yourself. Everyone has a self-image and self-descriptions, and sometimes the things you say about yourself can act as an obstacle to your safety commitment by restricting your actions. You might believe, "I am a safe person," or "I've never been hurt before, and I'm not going to get hurt today," and those ideas *can put you at risk*! Many of the words you use to describe yourself aren't always true and they aren't always helpful.

It's much more helpful to realize you are simply a unique individual who has experienced many things, but you are not *defined* by the words related to those experiences. In order to strengthen your commitments, it's important to let go of any unhelpful self-descriptions that divert you from acting safely. When you're unburdened from disruptive self-descriptions, you're more likely to follow through on what is important to you.

```
┌─────────────────────┐
│     Here Now        │
└─────────────────────┘
```

"Center your situational awareness on what you are doing in this work environment. Notice what is happening here and now and let go of distracting thoughts about events not in your present control."

Safety commitment requires action, and your actions can only happen *right now*. However, it's difficult to constantly pay attention to what is happening in the present moment. Sometimes we're not fully aware of what is going on around us while we are working. Our attention span is often lured away from a task-at-hand, and our mind often wanders into thinking about the past or the future rather than contacting the here and now. The fact that we are easily distracted can pose a threat while working at dynamic jobsites.

Building stronger situational awareness will help you focus on your job tasks, and be more in tune with the cues at your jobsite that remind you to act safely. Since work can only happen in the here and now, safety needs to be focused on here and now, too. It is critical to build up the skills that help you be alert to what is happening in both the external-world and internal-world right here and now.

"Allow yourself to acknowledge any emotions you are having without trying to control the emotions. Be willing to simply have those emotions while moving forward with safe and productive actions."

Like it or not, we experience an array of emotions during the workweek. Some emotions support our productivity and safety, but sometimes, we just don't "feel" like following through on our commitments. You can build up the personal skill of "accepting" to help you deal with unhelpful emotions. Accepting happens when you allow emotions to simply happen without trying to get rid of them or being hooked by them. It takes some practice, but you can learn to simply let them be without responding to them in a negative way.

For example, oftentimes, people want to get rid of feelings such as nervousness, so they walk away from things that are intimidating or scary. But if you walk away from fearful situations you should be facing (like giving feedback to a person doing the wrong thing, or telling your boss you need new PPE), you are letting your emotions get the best of you, and you aren't following through on your commitment to your safety.

Other times, emotions such as anger can cause you to act in a dangerous manner. You may lose control when you try to blow off steam, or perhaps you might act aggressively on the job because your emotions got you all steamed up. Either way, emotions will only have a negative impact on our behavior if we let them.

The healthy and safer way to deal with emotions is to accept the fact they will show up throughout your life, and simply observe that emotions are present without letting them push you into breaking your valued commitments. What you care about lasts a lifetime, but emotions last only a fraction of the day. Make sure the important things in your life guide your behavior on the job, and don't let your emotions interfere with your commitments. Accepting your feelings, and knowing they don't have to divert you from acting in the way you care about is an important skill to have.

Noticing

"Prepare to simply notice thoughts that arise during the work task. Let them go if they are not helpful. Treat distracting thoughts as disconnected from action while choosing to act in a meaningful and safe manner."

Your mind is a wonderful tool. It can generate dozens of thoughts every minute. Some thoughts are great for making decisions and solving problems. But sometimes those thoughts are interruptions to your committed work stream. Each of us can get lost in daydreaming about the future or reminiscing about the past, even while we're on the job. Often, we have unproductive thoughts such as, "This PPE is a waste of time," or "I'd finish this gig faster if I didn't have to go through all the safety precautions." Our own mind can act as a hazard to safety.

In Chapter 6, we discussed a skill called "noticing," which you can strengthen and apply to unsafe sidetracking thoughts. You can simply notice what your mind is throwing at you without actually giving into what your mind is saying. Just because your mind says something, doesn't mean you have to follow through if it isn't helpful. The exercises in Chapter 6 helped you have this experience yourself, and you can do one more very quickly right now: Hold your left hand above your head and think to yourself "I can't raise my left hand. No matter what I do, I can't raise my hand!"

You have the ability to disconnect the things you *say* in your internal-world from the things you *do* in the external-world. Noticing thoughts rather than being ruled by your thoughts is a helpful tool to use when your internal-world is taking you on a detour away from your committed actions.

Doing

For this section, you create your own To-Do list.

No matter what industry you work in, you're required to meet certain objectives at your job. Some of the objectives have to do with performing certain tasks, while others have to do with taking certain precautions. Listing these objectives in a way that makes it very obvious what you have to do at the worksite helps you maintain your commitment to optimum performance on the job. Following a checklist during the workday has many benefits, and when it comes to keeping commitments, To-Do lists provide measurable evidence you followed through on what you needed to do. No commitment is complete without action, therefore well-formulated To-Do lists should be in place to help you maintain your focus on behavior and productivity.

For this section, you fill in your own Safety Declaration section and Personal Values section. Describe why you work and why you personally aim for safety.

When you know why you work, and I mean *really know why* you work, then you have a very strong platform for choosing to act safely. It's difficult to commit to safety when you don't actually value safety. In Chapter 4, we discussed how some workers might go through their career simply *complying* with safety measures, and why compliance is not optimum safety. Compliance is about doing safety because of someone else's rules. Safety commitment is about doing safety because you actually care about it.

This book invited you to clarify your values and to truly uncover what is vital and significant in your life. Once you are clear on what is absolutely meaningful to you, you can choose to leverage those things as reminders to act safely! Sustaining an injury will hamper your ability to do the things you enjoy, and needless to say, death erases your ability to enjoy anything you've come to love in your life. When you know the critical, meaningful, and vital elements in your one life, then those ideas will personally motivate you to stick to your safety commitment.

There are two parts of the Values-Based Motivation section of the Safety Commitment Plan Worksheet: the Safety Declaration and the Personal Values section. In Chapter 3, we laid the foundation for a solid Safety Declaration that can be used by almost anyone: "Because my health, and the health of others, is important to me, I am willing to address external and internal-world obstacles that jeopardize safety, and to work in a manner that reflects safety as my top value." If that resonates with you, you can simply include it as part of the printed template, and read through it as a reminder every time you use this checklist. If you don't like that declaration, include your own personal replacement declaration you were encouraged to formulate in Chapter 3.

For the Personal Values section, I recommend you fill in the blanks with a pen or pencil every time you use this worksheet. Don't simply plug your values into the template, and take for granted those are the only things you care about. You can switch it up when you begin to take inventory on all the vital and meaningful things in your life. It's better to keep these motivations fresh, and write down different values as often as you use the checklist. That will help you approach the Safety Commitment Plan more sincerely, rather than simply pencil-whipping the boxes.

You can expand the template to include more rows to list more values. On the other hand, some people simply have one cardinal value that strengthens their safety commitment. Values are personal, so you should use this section in a way that generates the most motivation to take care of yourself.

Performance Management Contract

Describe incentives for complying with safety rules and meeting safety criteria.

Adding performance management principles to your Safety Commitment Plan keeps your behavior on-track because you are setting up reminders and rewards in the external-world to direct your safety actions. Signs, social influence, and accountability partners can channel your efforts to putting your best foot forward.

Performance management contracts were discussed in the Doing chapter, but it's more practical to write the contract at the bottom of the worksheet simply to maintain the flow of the cornerstone sentence. Performance management contracts, BBS processes, and accountability agreements sometimes take up a lot of room on the worksheet, so it makes sense to put it at the bottom. The performance management section of the Safety Commitment Plan should be developed by someone knowledgeable in incentive systems to ensure the right work actions are being rewarded.

Now that each building block has been established, you can compile the pieces to see that the Safety Commitment Plan Worksheet acts as a checklist. Your safety commitment will be increased when you use this tool before starting a work task. Step-by-step, row-by-row, you can prepare yourself to address your internal-world issues, such as distraction, frustration, or lack of direction, and then review the specific steps you'll take in the To-Do list, and be reminded of what you care about and why you will choose to act safely.

Safety Commitment Plan Worksheet

I am here now, accepting the way I feel,
noticing my thoughts, while doing what I care about.

	Addressing Internal-World Safety Issues	✔
I Am	Notice if you are being influenced by any unhelpful self-descriptions. Let go of anything unhelpful that you are believing about yourself.	
Here Now	Center your situational awareness on what you are doing in this work environment. Notice what is happening here and now and let go of distracting thoughts about events not in your present control.	
Accepting	Allow yourself to acknowledge any emotions you are having without trying to control the emotions. Be willing to simply have those emotions while moving forward with safe and productive actions.	
Noticing	Prepare to simply notice thoughts that arise during the work task. Let them go if they are not helpful. Treat distracting thoughts as disconnected from action while choosing to act in a meaningful and safe manner.	
Doing	**To-Do List for External-World Safety Actions**	
	Wear PPE	
	Communicate hazards immediately	
	Use tool guards	
	Perform lockout/tagout procedures	
What I Care About	**Values-Based Motivation**	
	Safety Declaration: Because my health, and the health of others, is important to me, I am willing to address external and internal-world obstacles that jeopardize safety, and to work in a manner that reflects safety as my top value.	
	--Personal Values: Describe why you work and why you aim for safety.	
	For my children	
	For my spouse	
	To continue making money to contribute to charity	

Performance Management Contract	
Signs and signals are in place	
Publicly announced safety commitment	
Accountability partner: _____ is aware of my commitments	

Describe incentives and performance criteria related to safety process:

Moving Forward

The core message of this book is to help you learn how to harness the powerful effect of personal safety commitment. Reviewing these principles on a daily basis, and truly incorporating the Safety Commitment Plan Worksheet can have a major impact on your dedication to your values, your ability to deal with internal-world obstacles, and most of all, your own external-world safety! Companies have modified these Safety Commitment Plan Worksheets to meet their own needs, and incorporated the process into the workday. The research behind using these concepts is well-founded (see Epilogue), and I hope you begin integrating these topics into your organization, or at least your own personal work process... right here and right now!

Epilogue

The Foundation for
Building Safety Commitment

"...I want you to do me a favor...
commit yourself to doing honest work."

At the beginning of this book I told you my story about promising to commit to doing honest work. This book about commitment is based on that promise of honesty. The research-based principles in this book are aimed to help you develop your Safety Commitment Plan, and come from a proven approach for improving behavior.

The six-point model used throughout this book to help you build safety commitment comes from an area of applied behavioral science called Acceptance and Commitment Training (ACTraining). You can see the word "commitment" in the title of that scientific protocol. When used in psychology clinics, this technology is called Acceptance and Commitment Therapy (ACT), and ACT has been shown to be a highly effective way to support behavioral change. ACT has assisted people who struggle with depression, anxiety, and drug addiction in keeping their commitments. In addition to mental health issues, research has shown that ACTraining is effective for improving performance and other important variables in the workplace.

You are likely already convinced about the effectiveness of using To-Do lists and performance management because these tools are typically used in industries where performance is important. However, the other principles in this book might seem foreign to many industrial settings. In my years as a construction worker, I can't recall my coworkers talking about values or emotions. But despite the fact that these topics are unconventional, I hope you participated in the exercises throughout this book, and were able to understand the personal relevance of values-based motivation, acceptance of emotions, noticing thoughts, having a clear sense of self, and being in the present moment.

Incorporating mindfulness and situational awareness in the workplace is becoming more broadly embraced. The *Wall Street Journal*, *Harvard Business Review*, and the *New York Times* have all published stories about the benefits of mindfulness in the workplace. There is a good deal of research backing the application of this model and the Additional Readings section of this book has a list of articles demonstrating the effectiveness of the approach.

The aim of this book is to integrate new scientific findings with what has been traditionally successful in helping improve industrial safety. The effectiveness of ACTraining is the foundation of *Building Safety Commitment*. I hope you see the merit of these principles, incorporate them into your habits, and face each workday able to say, "I am here now, accepting the way I feel, noticing my thoughts, while doing what I care about" so you can continue acting in the direction of what you care about even in the presence of obstacles. In other words, I hope this book helps you in building safety commitment.

Endnotes

1 – Solis, Hilda L. (2012). Remarks from the Honorable Hilda L. Solis "Workers Memorial Day" United States Department of Labor, Los Angeles, CA. http://www.dol.gov/_sec/media/speeches/20120426_WMD.htm

2 – McSween, T. E. (2003). *Values-based safety process: Improving your safety culture with behavior-based safety.* Wiley-Interscience, Hoboken, NJ.

3 – Bond, F.W. & Bunce, D. (2000). Mediators of change in emotion-focused and problem-focused worksite stress management interventions. *Journal of Occupational Health Psychology, 5* (1), 156-163.

4 – Bond, F.W. & Bunce, D. (2003). The role of acceptance and job control in mental health, job satisfaction, and work performance. *Journal of Applied Psychology, 88* (6), 1057-1067.

5 – Bond, F.W. & Flaxman, P.E. (2006). The ability of psychological flexibility and job control to predict learning, job

performance, and mental health. *Journal of Organizational Behavior Management, 26* (1), 113-130.

6 – Bond, F.W., Flaxman, P.E, & Bunce, D. (2008). The influence of psychological flexibility of work redesign: Mediated moderation of a work reorganization intervention. Journal of Applied Psychology, 93 (3), 645-654.

7 – Luoma, J. B., Hayes, S. C., Twohig, M. P., Roget, N., Fisher, G., Padilla, M., Bissett, R., & Kohlenberg, B. (2007). Augmenting continuing education with psychologically focused group consultation: Effects on adoption of group drug counseling. *Psychotherapy: Theory, Research, Practice, Training, 44*(4), 463-469.

8 – Carmona, R. H. (2004). *The health consequences of smoking: A report of the surgeon general.* U.S. Department of Health & Human Services, Washington, DC.

9 – Office of National Drug Control Policy (2004). *The economic costs of drug abuse in the United States.* Washington, DC.

10 – Harwood, H. (2000). *Updating estimates of the economic costs of alcohol abuse in the United States: Estimates, update methods, and data,* Report prepared by the Lewin Group for the National Institutes on Alcohol Abuse and Alcoholism, Washington, DC.

11 – Substance Abuse and Mental Health Services Administration (1997). *An analysis of worker drug use and workplace policies and programs.* OAS Analytic Series #A-2, DHHS Publication No. (SMA) 97-3142, Rockville, MD.

12 – U.S. Department of Justice (1994). *Work place violence.* Washington, DC.

13 – Bureau of Labor Statistics (2012). *National census of fatal occupational injuries in 2011.* Washington, DC.

14 – Killingsworth, M. A. & Gilbert, D. T. (2010). *A wandering mind is an unhappy mind.* Science (303), 931-932.

15 – Smith, K., & Hancock, P.A. (1995). Situation awareness is adaptive, externally directed consciousness. Human Factors, 37 (1), 137–148.

16 – Adam, E.C. (1993). Fighter cockpits of the future. Proceedings of 12th IEEE/AIAA Digital Avionics Systems Conference (DASC), 318–323.

17 – Gawande, A. (2009). *The checklist manifesto: How to get things right.* Holt & Company, NY.

18 – Degani, A. & Wiener, E. L. (1993). *Cockpit checklists: Concepts, design, and use.* Human Factors 35 (2), 28-43.

19 – Pronovost, P. & Vohr, E. (2011). Safe patients, smart hospitals: How one doctor's checklist can help us change health care from the inside out. Plume Publishing/ Penguin: NY.

20 – Occupational Safety and Health Administration (2002). *Job hazard analysis*. U.S. Department of Labor, Washington, DC.

21 – Roughton, J. E. & Crutchfield, N. (2007). Job hazard analysis: A guide for voluntary compliance and beyond. Elsevier: MA.

22 – Medina, R.E, McSween, T. E., Rost, K. & Alvero, A. M. (2009). *Behavioral safety in a refinery*. Professional Safety, (August), 36-40.

23 – McSween, T. E. (2003). *Values-based safety process: Improving your safety culture with behavior-based safety*. Wiley-Interscience, Hoboken, NJ.

Additional Readings

Atkins, P.W.B. & Parker, S.K. (2012). Understanding individual compassion in organizations: The role of appraisals and psychological flexibility. *Academy of Management Review*, 37 (4), 524-546.

Bond, F.W., Hayes, S.C., & Barnes-Holmes, D. (2006). Psychological flexibility, ACT, and organizational behavior. *Journal of Organizational Behavior Management*, 26 (1), 25-54.

Bond, F. W., Lloyd, J. & Guenole, N. (2012). The work-related acceptance and action questionnaire (WAAQ): Initial psychometric findings and their implications for measuring psychological flexibility in specific contexts. *Journal of Occupational and Organizational Psychology*, 1-25.

George, B. (2012). Mindfulness helps you become a better leader. Harvard Business Review Blog Network. http://blogs.hbr.org/hbsfaculty/2012/10/mindfulness-helps-you-become-a.html

Moran, D. J. (2010). ACT for leadership: Using acceptance and commitment training to develop crisis-resilient change managers. *International Journal of Behavioral Consultation and Therapy, 6 (4)*, 341-355.

Varra, A.A., Hayes, S.C., Roger, N. & Fisher, G. (2005). A randomized control trial examining the effect of acceptance and commitment training on clinician willingness to use evidence-based pharmacotherapy. *Journal of Consulting and Clinical Psychology, 76* (3), 449-453.

Appendix A

Use the Safety Commitment Plan Worksheet in order to create a personal approach to optimizing your own safety. When you visit www.buildingsafetycommitment.com, you will be able to download Safety Commitment Plan Worksheets in PDF.

Safety Commitment Plan Worksheet

I am here now, accepting the way I feel,
noticing my thoughts, while doing what I care about.

	Addressing Internal-World Safety Issues	✔
I Am		
Here Now		
Accepting		
Noticing		
Doing	**To-Do List for External-World Safety Actions**	
What I Care About	**Values-Based Motivation**	

Performance Management Contract

Signs and signals are in place	
Publicly announced safety commitment	
Accountability partner: _____ is aware of my commitments	

Describe incentives and performance criteria related to safety process:

Appendix B

In an effort to strengthen your safety commitment, Quality Safety Edge & Valued Living Books has created a website with resources aimed to help you put this book to work for you. When you visit www.buildingsafetycommitment.com, you will be able to download Safety Commitment Plan Worksheets in PDF, and also find audio recordings of the exercises found in this book. These resources are free and highly recommended if you are interested in increasing your situational awareness, acceptance, and noticing skills. Just make sure that you are in a safe and comfortable environment while you listen to these exercises!

Notes